THE AI OWNER'S MANUAL
A Practical Guide for Business Owners

By Jess Coburn

Copyright © 2026 Jess Coburn

DISCLAIMER

The information contained in this book is for general informational purposes only. While we have made every effort to ensure the accuracy and completeness of the information presented, the rapidly evolving nature of artificial intelligence means that specific tools, capabilities, pricing, and best practices may change after publication.

The opinions expressed in this book are those of the author and do not necessarily reflect the views of any AI vendors, technology companies, or other third parties mentioned herein. Product names, logos, and brands are the property of their respective owners and are used for identification purposes only.

This book does not constitute legal, financial, or professional advice. Readers should consult with qualified professionals regarding their specific circumstances before making business decisions based on the information provided.

NO WARRANTY

The author and publisher make no representations or warranties with respect to the accuracy, applicability, or completeness of the contents of this book. The author and publisher specifically disclaim any implied warranties of merchantability or fitness for a particular purpose. Neither the author nor the publisher shall be liable for any loss of profit or any other commercial damages, including but not limited to special, incidental, consequential, or other damages.

THIRD-PARTY CONTENT

All product names, logos, services and brands are property of their respective owners. All company, product, and service names used in this book are for identification purposes only. Use of these names, logos, and brands does not imply endorsement.. Pricing, features, and availability mentioned were accurate at the time of writing but are subject to change. Readers should verify current information directly with vendors.

First Edition: 2026

ISBN: 9798905141362

Preface

Over the past few years, I've had the privilege of working with hundreds of business leaders. CEOs, CFOs, COOs, and department heads at companies ranging from twenty employees to five hundred. These conversations have happened in boardrooms and coffee shops, on video calls and factory floors. And no matter the industry, company size, or individual background, I kept hearing the same concerns.

"I know AI is important, but I don't know where to start."

"Every vendor promises the world, but how do I know what's real?"

"I don't want to fall behind, but I also can't afford expensive mistakes."

These weren't technical questions. They were business questions. And they deserved practical, honest answers-not the breathless hype that dominates most AI conversations.

That's why I wrote this book.

The Problem with Most AI Advice

Most books about AI fall into one of two camps. The first is written by technologists for technologists. They're full of jargon, algorithms, and implementation details that assume you want to become a machine learning engineer. The second is written by futurists and consultants. Painting grand visions of AI-transformed industries while offering little guidance on what to do Monday morning.

Neither serves the executive who needs to make real decisions with real budgets and real teams.

Throughout my career, I've built my reputation on cutting through complexity to deliver practical technology solutions. The business leaders I've worked with don't want theory, they want results. And that's exactly what I've tried to capture in these pages.

What You'll Find in This Book

This book is organized around the questions that keep executives up at night. We start with a reality check.

Separating what AI can actually do today from the marketing hype. Then we move through a logical progression: finding your starting point, implementing your first wins, building governance, managing change, and ultimately creating sustainable competitive advantage.

Every chapter includes frameworks you can apply immediately, questions you should be asking, and common pitfalls to avoid. I've drawn from real implementations (some successes and some failures) to give you guidance that's been tested in the field.

What matters more: I've tried to be brutally honest. I'll tell you when something isn't ready for prime time. I'll point out when the emperor has no clothes. I'll acknowledge the legitimate concerns about AI's impact on jobs and society. You won't find cheerleading here. Just practical guidance from a guy leading a team that implements this technology every day.

A Note on Timing

AI is evolving at an extraordinary pace. Between the time I finish writing this sentence and the time you read it, new capabilities will have emerged, new tools will have launched, and some of the specific recommendations in this book may need updating.

That's why I've focused on principles rather than products, frameworks rather than features through the first 10 chapters. The specific tools will change; the fundamental approach to evaluating and implementing them won't.

After chapter 10 I've added a few more chapters on innovative technology we're helping companies like yours implement today. But those technologies are changing at light speed and by the team you read this, the ball has probably already moved.

That's why I've created an online companion resource at https://jesscoburn.com/executive-ai/ where you can access many of the updated templates, calculators, checklists, and reference guides. The supplemental tools mentioned throughout this book and many more are available for download there, along with updates as the market evolves.

How to Use This Book

You can certainly read this book cover to cover, and the chapters do build on each other. But I've also designed it as a practical reference you can return to as needs arise. Evaluating a vendor? Turn to Chapter One's BS Detector framework. Planning your first ninety days? Chapter Three has your roadmap. Trying to get buy-in from a skeptical team? Chapter Eight addresses change management head-on. These tools and dozens more are available on the accompanying website: https://jesscoburn.com/executive-ai.

Each chapter ends with a forty-eight-hour challenge. This is a concrete action you can take immediately to apply what you've learned. Don't skip these. The gap between knowing and doing is where most AI initiatives die.

The businesses that will thrive in the coming decade are the ones that figure out how to use AI as a tool for human amplification, not replacement. The ones that start with business problems rather than technology solutions. The ones that build cultures of continuous learning rather than one-time implementations.

You've already taken the first step by picking up this book. Now let's get to work.

Jess Coburn

May 2026

Acknowledgments

No book is a solo effort. This one grew out of conversations, implementations, and failures.

To the businesses that trusted me and my team with their technology journeys over the last 27 years, and then their AI journeys too: you shaped everything here by proving that the best solutions start with the business problem, not the technology.

Finally, to my family, for the patience and the perspective.

The work continues.

Table of Contents

THE AI OWNER'S MANUAL

A Practical Guide for Business Owners

THE AI OWNER'S MANUAL

The Executive's AI Reality Check

What AI can actually do for your business today- not in some theoretical future

Module Overview

If you've been following AI news, you've probably experienced a particular kind of whiplash. One day you read that AI will replace half of all jobs within five years. The next day you hear it can't reliably count the letters in a word. Headlines swing between "AI will solve climate change" and "AI chatbots confidently make up facts." It's enough to make any sensible business leader wonder: what's actually true?

This module cuts through both the hype and the skepticism to give you a clear-eyed view of AI's actual capabilities in early 2026. You'll learn to distinguish between what AI does extraordinarily well, where it struggles, and how to evaluate vendor claims without needing a computer science degree. By the end, you'll have a practical framework for understanding any AI pitch that comes your way.

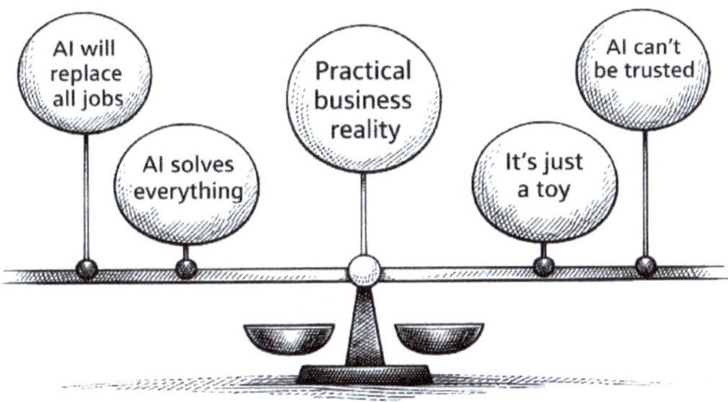

Learning Objectives

- ✓ Understand the three categories of AI applications: Assistants, Automations, and Agents
- ✓ Recognize where AI delivers real ROI versus where it's oversold
- ✓ Apply the "3 Questions" framework to evaluate any AI vendor claim
- ✓ Know the specific limitations that trip up most organizations

the AI space in Plain English

Before we look at what AI can do for your business, let's establish a shared vocabulary. The term "AI" gets thrown around to describe everything from spam filters to robots that flip burgers, which makes productive conversations nearly impossible. Here's what matters for business leaders in 2026.

What "AI" Actually Means Today

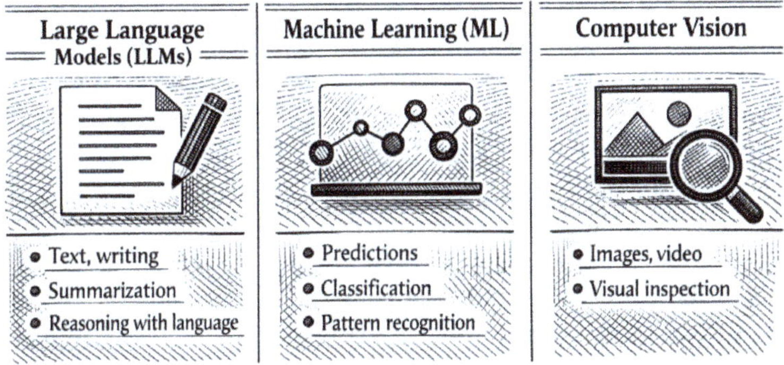

When vendors and media talk about AI in 2026, they're usually referring to one of three things, and the differences matter enormously for your business decisions.

Large Language Models (LLMs) are the technology behind ChatGPT, Claude, Gemini, Grok, and similar tools. These systems excel at understanding and generating human language. They perform tasks like summarizing documents, drafting emails, answering questions, and writing code with ease and accuracy. Think of them as incredibly well-read assistants who can process and generate text at superhuman speed. The leading models in early 2026 include Claude Opus from Anthropic, GPT-5.X from OpenAI, and Gemini 3 from Google. Each has slightly different strengths, but for most business applications, they are remarkably capable across the board.

Machine Learning (ML) refers to systems that learn patterns from data to make predictions or decisions. Your bank's fraud detection, Netflix's recommendations, and your email's spam filter all use ML. These systems are narrower than LLMs, they do one thing well, but they have been quietly revolutionizing businesses for over a decade. When a vendor talks about "predictive analytics" or "intelligent forecasting," they typically mean ML.

Computer Vision enables machines to interpret images and video. This powers everything from quality control cameras on manufacturing lines to the facial recognition that unlocks your phone. For most small and medium businesses, computer vision applications are still fairly specialized, though that is changing rapidly.

For the purposes of this course, we'll focus primarily on LLMs and how they're being deployed in business contexts, since that's where most immediate opportunities exist for your business. The barrier to entry is low (often just a $20/month subscription), the applications are broad, and you don't need specialized technical staff to get started.

The Three Categories of AI Applications

The Three Categories of AI Applications

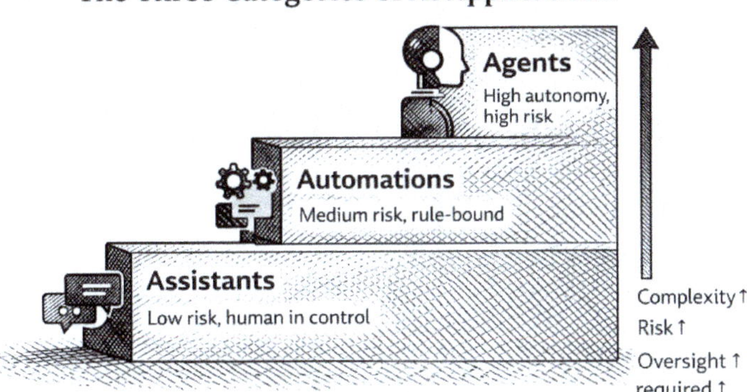

Every AI application your business might consider falls into one of three categories, each with different complexity, cost, and risk profiles. Understanding these categories helps you evaluate opportunities and set expectations.

Category	Description	Complexity	Best For
Assistants	AI helps humans work faster	Low	Most SMBs should start here
Automations	AI handles routine tasks	Medium	Mature processes with clear rules
Agents	AI pursues goals autonomously	High	Observe for now; approach with caution

Assistants are AI tools that help humans work faster and better. You ask a question, it provides an answer. You give it a draft, it suggests improvements. The human stays in control, and the AI's role is to accelerate what you're already doing. Examples include using ChatGPT to draft emails, Claude to summarize long documents, or Copilot to help write code. This is where most SMBs should start: low risk, immediate value, minimal setup. The human reviews everything before it goes out the door, which catches AI mistakes and maintains quality.

Automations are AI systems that handle routine tasks without human intervention for each instance. Think of email auto-responses, automatic invoice processing, or chatbots answering FAQs. These offer significant time savings but require more upfront setup and careful boundaries. The key question with automations is: what happens when the AI encounters something it wasn't designed for? Good automations include clear escalation paths to humans. Bad ones leave customers frustrated or create costly errors.

Agents are the newest and most ambitious category: AI systems that can pursue goals across multiple steps, make decisions, and take actions in the real world. An AI agent might research a topic, draft a report, schedule a meeting to discuss it, and send follow-up emails, all from a single instruction. In early 2026, agent technology is advancing rapidly but remains the highest-risk category. Agents can accomplish remarkable things, but they can also go spectacularly wrong when they misunderstand objectives or encounter situations their designers didn't anticipate. Most SMBs should observe this space rather than deploy agents in high-stakes contexts without the assistance of a technology provider.

The practical implication: start with assistants, graduate to automations where they make sense, and approach agents with healthy caution. Each step up the ladder offers greater efficiency but requires more sophisticated oversight.

Where AI Actually Delivers ROI

Where AI Actually Delivers ROI

Documents Email Meetings

Notes Research

"Small time savings, repeated daily, compound fast."

Flashy AI projects →   Boring productivity AI →

High effort, uncertain payoff   *Low effort, reliable payoff*

Let's get specific about where AI is delivering real, measurable value for businesses similar to yours. These aren't theoretical possibilities, they're applications that SMBs are using profitably right now.

The "Boring AI" That Actually Works

The most valuable AI applications aren't the flashy ones that make headlines. They're mundane productivity improvements that save 15 minutes here, an hour there, across tasks your team does every day. Cumulatively, these "boring" applications often deliver more ROI than ambitious transformation projects.

Document Processing and Summarization

This is likely where AI delivers the most immediate, consistent value. Modern language models can read a 50-page contract and extract the key terms in minutes. They can summarize customer feedback from hundreds of survey responses, identify patterns in support tickets, or distill a dense research report into actionable insights. The time savings are substantial as tasks that took hours now take minutes and the quality is genuinely good as long as humans spot-check the outputs.

Realistic ROI: A professional who spends 5 hours per week reading and summarizing documents can typically cut that to 1-2 hours with AI assistance, while actually covering more material. At a loaded cost of $75/hour, that's roughly $15,000-20,000 in annual productivity gains per person.

First-Draft Writing

AI excels at generating solid first drafts of routine business communications: emails, reports, proposals, job postings, meeting agendas. The key word is "first draft" because you'll still need to edit and personalize the output. But starting with a competent draft rather than a blank page speed ups the writing process. Some executives report cutting their email drafting time by 50-70%.

Realistic ROI: An executive who spends 10 hours per week on written communications can typically save 3–5 hours with AI-assisted drafting. What matters more: the reduction in cognitive load is meaningful. Let's face it, less time staring at a blank page means more mental energy for decisions that actually require human judgment.

Research and Information Synthesis

Need to understand a new market, evaluate a vendor, or prepare for a meeting with an unfamiliar prospect? AI tools can synthesize publicly available information in minutes rather than hours. They're particularly good at generating questions you should ask, identifying considerations you might have missed, and structuring your thinking about complex topics.

Important caveat: AI models have knowledge cutoffs and can occasionally get facts wrong. For critical decisions, always verify key claims from authoritative sources. Use AI to accelerate your research, not replace it entirely.

Meeting Preparation and Follow-Up
From generating agendas to synthesizing meeting notes into action items, AI can handle the administrative work that surrounds meetings. Several tools now offer real-time transcription and summary, though quality varies. The clear wins are in post-meeting processing: turning rambling discussions into structured notes with clear decisions and assigned actions.

Code Generation and Technical Documentation
If your business employs developers, AI coding assistants are delivering substantial productivity gains with businesses frequently reporting 20-40% faster development for routine tasks. Beyond writing code, these tools excel at explaining existing code, generating documentation, and helping developers learn new technologies. Even if you're not a technical business, AI can help create and maintain the technical documentation that often gets neglected.

Where AI Falls Short

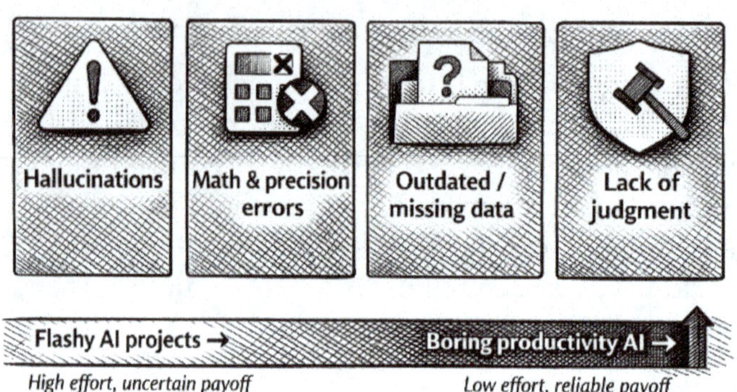

An honest assessment of AI limitations is essential for setting realistic expectations and avoiding expensive mistakes. Here's where AI consistently underperforms the marketing hype.

The Hallucination Problem

AI models sometimes generate confident sounding information that is completely fabricated. They'll cite nonexistent research papers, invent statistics, or describe events that never happened. All with the same authoritative tone they use for accurate information. This "hallucination" problem has improved significantly with newer models, but it hasn't been eliminated. Any AI output that includes specific facts, figures, or citations requires human verification.

Practical implication: Never let AI outputs that contain factual claims go directly to customers, partners, or legal documents without human review. The reputational and legal risks of publishing AI-generated misinformation far outweigh the time saved by skipping review.

Complex Reasoning and Math

Despite their impressive language abilities, current AI models struggle with complex mathematical reasoning, multi-step logic problems, and tasks requiring precise counting or spatial reasoning. They can make arithmetic errors, lose track of complex logical chains, and fail at tasks that seem simple to humans (like counting how many times a letter appears in a word). This is improving as the latest models are markedly better than those from even a year ago, but it remains a limitation.

Practical implication: Do not rely on AI for financial calculations, statistical analysis, or any task where precision matters. Use spreadsheets or specialized software for numbers and use AI for the narrative that wraps around those numbers.

Real-Time and Proprietary Information

AI models are trained on data up to a certain cutoff date and don't inherently know about events after that date or information they weren't trained on. They don't have access to your company's internal databases, your industry's private benchmarks, or yesterday's news unless you explicitly provide that information. Some tools now include web search capabilities, but these add latency and their own reliability challenges.

Practical implication: For questions requiring current information or company-specific data, either use AI tools with verified search capabilities or plan to supplement AI outputs with your own research.

Judgment and Context

AI doesn't understand your business context, your relationships, or the political dynamics of your organization. It can't tell when a technically correct answer would be politically disastrous, when a customer's complaint signals a deeper relationship problem, or when a "good" solution violates your company's unwritten rules. It also lacks the judgment to know when a task is inappropriate or when it should push back on instructions.

Practical implication: AI is a tool, not a decision-maker. It can inform decisions by providing analysis and options, but the judgment of which option to pursue, when to make exceptions, and how to handle complex human dynamics remains firmly human territory.

The "3 Questions" Framework for Evaluating AI Claims

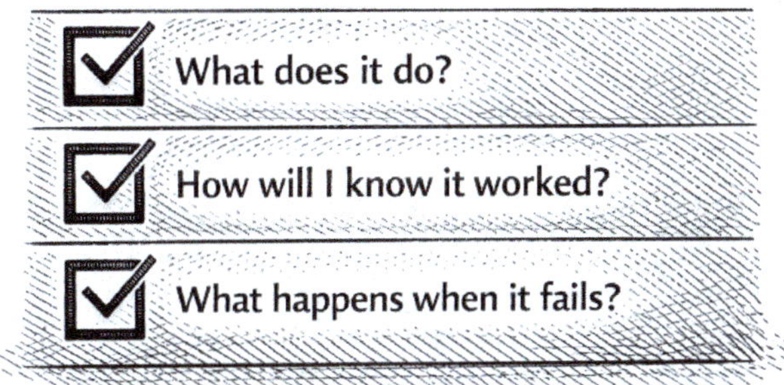

Every week brings new AI products promising to transform your business. Most won't deliver on their marketing claims. Use these three questions to quickly separate substance from hype.

Question 1: What Specifically Does It Do?

Vague claims like "uses AI to improve productivity" or "AI-powered insights" are red flags. Push for specifics: What exact task does it perform? What inputs does it need? What outputs does it produce?

Good answer: "Our tool reads your sales emails and automatically extracts action items, deadlines, and customer commitments into your CRM."

Red flag answer: "Our AI platform leverages machine learning to optimize your business processes."

Question 2: How Will I Know It Worked?

Any legitimate AI application should have measurable success criteria. If the vendor can't tell you how to measure whether their product is working, that's a problem.

Good answer: "Track time spent on data entry before and after. Our customers typically see 60-70% reduction within 30 days."

Red flag answer: "You'll just feel more productive" or "The AI continuously improves your results."

Question 3: What Happens When It Fails?

All AI systems fail sometimes. What matters is how failures are handled. What's the error rate? How will you know when it makes a mistake? What's the recovery process?

Good answer: "Our system flags low-confidence outputs for human review. About 15% of items need manual verification, and we provide a daily exception report."

Red flag answer: "Our AI is 99% accurate" (without explaining what happens in the other 1%) or "It learns from mistakes automatically."

These questions aren't designed to be gotchas, they're the same questions any competent vendor should have clear answers to. If a salesperson struggles to answer them, it usually means the product isn't mature enough for business use, or they don't understand what they're selling.

Evaluating AI Vendors: What to Look For

Beyond the 3 Questions framework, here are specific factors to evaluate when considering any AI product or service.

Data Security and Privacy

This is non-negotiable. Before using any AI tool with business data, understand where your data goes, how your data is stored, and whether your data is used to train the AI model. Remember, it's your data and today, data is king. You don't want to give it away freely.

Key questions to ask include: Is data encrypted in transit and at rest? Can you opt out of training data usage? What's their data retention policy? Do they have SOC 2 certification or equivalent? For enterprise tools, also ask about data residency options if you have geographic requirements.

Integration Capabilities

An AI tool that doesn't connect to your existing systems creates friction that often kills adoption. Ask about native integrations with your key platforms (CRM, email, project management), API availability for custom integrations, and single sign-on support. The best tool in the world is useless if your team won't use it because it's isolated from their workflow.

Look for vendors that support MCP (Model Context Protocol), an modern standard that lets AI tools connect to your existing software in standardized ways. This is becoming table stakes for enterprise AI tools and significantly simplifies integration compared to custom API work.

Pricing Transparency

AI pricing models vary wildly: per-user, per-query, per-document, usage-based with complex tiers. Understand exactly how you'll be charged and what happens if your usage grows. Ask for typical customer cost ranges and whether there are overage charges. Some "cheap" per-seat licenses become expensive when you account for usage limits.

Support and Reliability

What's the vendor's uptime history? What support is included? Is there a service level agreement (SLA)? For mission-critical applications, you need vendors with proven reliability and responsive support. Check their status page history if available and ask for customer references you can contact.

Company Viability

The AI market is flooded with startups, and not all will survive. Before betting on a vendor, consider their funding and runway, their customer base size and growth, and the team's track record. This doesn't mean avoiding startups as some of the best AI tools come from newer companies. But you should factor company risk into your evaluation, especially for applications where switching costs would be high.

What This Looks Like in Practice

Abstract principles are helpful, but concrete examples make them real. Here's how businesses similar to yours are actually using AI today; the specific applications, the realistic results, and the lessons learned.

Professional Services Firm (45 employees)

Application: Document summarization and proposal drafting

Tools used: Claude Pro subscriptions for professionals, with enterprise governance policies

Results: Reduced proposal development time by 40%. Senior staff estimate saving 5-8 hours per week on document review and drafting.

Key lesson: Started with an AI acceptable use policy before rolling out tools. The upfront governance work prevented data security incidents and helped reluctant partners feel comfortable with the technology.

E-commerce Company (120 employees)

Application: Customer service response assistance and product description generation

Tools used: Custom solution built on GPT-5.2 API integrated with their helpdesk and e-commerce platform

Results: Customer service team handles 30% more tickets per day. Product team creates new listings 3x faster. Customer satisfaction scores maintained (no decline).

Key lesson: Spent three months piloting with a small team before company-wide rollout. Discovered and fixed several edge cases where AI suggestions were inappropriate before they reached customers.

Manufacturing Company (80 employees)

Application: Technical documentation and procedure writing

Tools used: Microsoft Copilot integrated with SharePoint and existing documentation systems

Results: Created thorough procedure documentation for 40 processes that had never been documented. Estimated time savings: 6 months of equivalent manual documentation effort.

Key lesson: The value wasn't just time savings, it was finally getting documentation done that had been perpetually deprioritized. AI lowered the barrier enough that documentation actually happened.

Where It Didn't Work

Equally instructive are the failures. Common patterns we've observed include organizations that tried to use AI for customer-facing communications without human review (resulting in embarrassing mistakes that damaged client relationships), companies that purchased expensive AI platforms before identifying specific use cases (resulting in shelfware), and teams that deployed AI automation without clear escalation paths (resulting in frustrated customers stuck in loops).

The common thread in failures: skipping foundational work (governance, use case definition, human oversight) in the rush to "do AI."

Key Takeaways

Let's distill this module into actionable principles you can apply immediately.

1. **AI is a tool, not magic.** Current AI excels at specific tasks (language processing, pattern recognition) while

steadily improving with others (complex reasoning, real-time information). Match your applications to your AI platform's actual strengths.

2. **Start with assistants.** The most reliable AI value comes from augmenting human work rather than replacing it. AI that helps your team work faster is lower risk and often higher ROI than AI that works autonomously.

3. **Human oversight is essential.** AI makes mistakes, sometimes subtle ones. Until your team develops an intuition for where AI fails, review everything before it goes external.

4. **"Boring" applications often have the best ROI.** Document summarization, draft writing, research assistance and similar mundane applications compound into significant productivity gains.

5. **Use the 3 Questions framework.** When evaluating any AI opportunity: What specifically does it do? How will I know it worked? What happens when it fails?

6. **Governance before tools.** Establish your AI acceptable use policy before rolling out AI applications. This protects your company and actually accelerates adoption by reducing uncertainty.

Your 48-Hour Challenge

This course is designed for action, not just learning. Before moving to Module 2, complete at least one of these exercises:

1. **Test an AI assistant:** Sign up for a free tier of ChatGPT, Claude, or Gemini. Use it to summarize a long document you've been putting off reading. Note how long it takes and whether the summary captures the key points.
2. **Apply the 3 Questions:** Think of an AI vendor pitch you've received recently (or search for one in your inbox). Run it through the 3 Questions framework. Does it hold up?
3. **Identify your "boring" opportunities:** List three routine tasks you or your team do repeatedly that involve reading, writing, or summarizing. These are likely your highest-ROI AI opportunities.
4.

What's Next

If you remember nothing else from this module, remember the 3 Questions framework. Write them on a sticky note if you have to. Every vendor pitch, every LinkedIn post about some new AI tool, every "you need this" recommendation from a consultant should run through those three filters before you spend a dollar or an hour of your team's time.

Now that you know what AI actually is (and isn't), the next question is obvious: where do you start? Module 2 gives you a scoring framework to figure that out.

THE AI OWNER'S MANUAL

AI Triage: Finding Your Starting Point

Walk away knowing exactly which process in your business to tackle first with AI

Module Overview

You now understand what AI can do. You're past the hype, you have realistic expectations, and you're ready to act. But here's where many executives get stuck: the paralysis of too many possibilities.

Should you start with customer service? Operations? Finance? Your marketing team is begging for AI tools, but so is your sales manager. Every vendor demo looks promising. Every use case seems viable. And because everything could benefit from AI, you end up doing nothing. Or worse, doing everything at once and watching it all collapse under its own complexity.

This module solves that problem. You'll learn a systematic framework for identifying and prioritizing AI opportunities specific to your business. By the end, you won't just have a vague sense of where to start. You'll have a ranked list of specific processes with clear reasoning for why each one does or doesn't make sense right now.

The goal isn't to find every AI opportunity. It's to find the right first one: the project that builds momentum, demonstrates value, and creates the foundation for everything that comes after.

Learning Objectives

- √ Apply the Impact/Effort/Risk framework to evaluate any AI opportunity.
- √ Identify which processes in your business are best suited for AI
- √ Use "process mining lite" techniques to find hidden opportunities.
- √ Balance quick wins against strategic plays for sustainable AI adoption
- √ Create a prioritized AI opportunity list specific to your business.
- √ Know the specific limitations that trip up most organizations

The AI Opportunity Framework

Not all AI opportunities are created equal. Some look exciting but consume enormous resources. Others seem boring but deliver reliable returns. The key is having a systematic way to evaluate and compare opportunities so you're making decisions based on analysis, not enthusiasm.

We use a three-dimensional framework: **Impact**, **Effort**, and **Risk**. Each dimension captures something essential about whether an AI opportunity makes sense for your business right now.

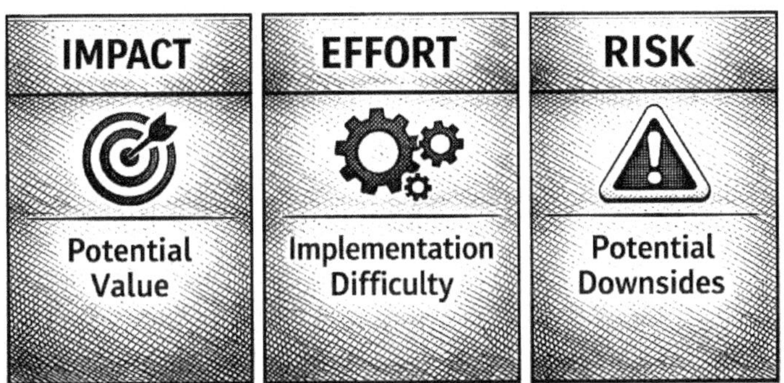

Dimension 1: Impact

Impact measures the potential value if this AI application succeeds. We assess impact across four sub-dimensions:

Time Savings: How many person-hours per week/month would this save? Be specific. "Saves time" isn't good enough. Estimate actual hours. A task that takes 2 hours weekly for 5 people is 40 hours monthly (or roughly a quarter of an FTE).

Quality Improvement: Would AI improve the consistency or quality of outputs? This matters especially for customer-facing work where errors are costly. Consider both direct quality gains and reduced variance.

Strategic Value: Does this unlock capabilities you couldn't have otherwise? Some AI applications enable things that were simply impossible before. Like personalizing communications at scale or analyzing patterns across thousands of documents.

Revenue/Cost Effect: Can you tie this to dollars? Time savings convert to labor costs. Quality improvements reduce rework and customer churn. Strategic capabilities might enable new revenue streams or pricing power.

Impact Scoring Guide

Score	Criteria
1 (Low)	Minor time savings (<5 hours/month across team), no quality or strategic benefit
2	Moderate time savings (5-20 hours/month), some quality benefits
3	Significant time savings (20-50 hours/month), clear quality improvements
4	Major time savings (50+ hours/month) OR meaningful revenue/cost impact
5 (High)	Transformative: enables new capabilities, significant competitive advantage, or major financial impact

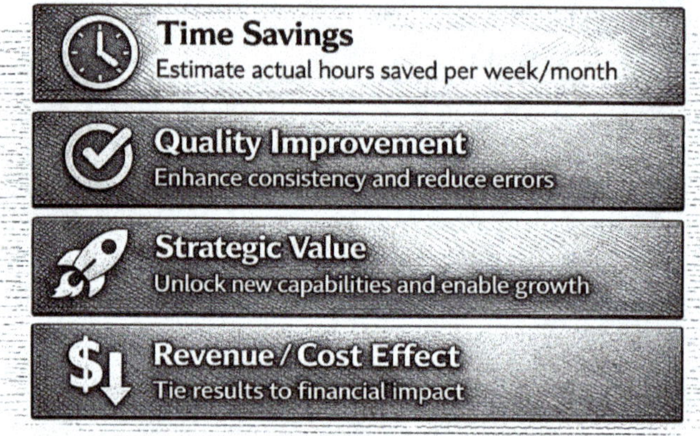

Dimension 2: Effort

Effort measures what it takes to implement and maintain the AI application. Low effort doesn't mean low value. Often the easiest wins are the best ones. Assess effort across these sub-dimensions:

Technical Complexity: Does this require custom development, or can you use existing tools? A ChatGPT subscription plus training is vastly different from building a custom model integrated with your ERP.

Data Requirements: What data does the AI need? Is that data accessible, clean, and in usable format? Data preparation is often 80% of AI project effort.

Change Management: How much will this change how people work? A tool that fits into existing workflows is easier than one requiring new processes. Consider who needs to adopt it and their likely resistance.

Ongoing Maintenance: What does it take to keep this running? Prompts need refinement, models may need retraining, processes need monitoring. Factor in the long-term commitment, not just the initial setup.

Effort Scoring Guide

Score	Criteria
1 (Easy)	Off-the-shelf tool, minimal config, fits existing workflow, 1-2 days to implement
2	Standard tool with customization, some workflow changes, 1-2 weeks
3	Integration required, moderate process changes, some data preparation, 1-2 months
4	Custom development, significant data work, major process changes, 3-6 months
5 (Hard)	Complex custom solution, extensive data infrastructure, organizational transformation, 6+ months

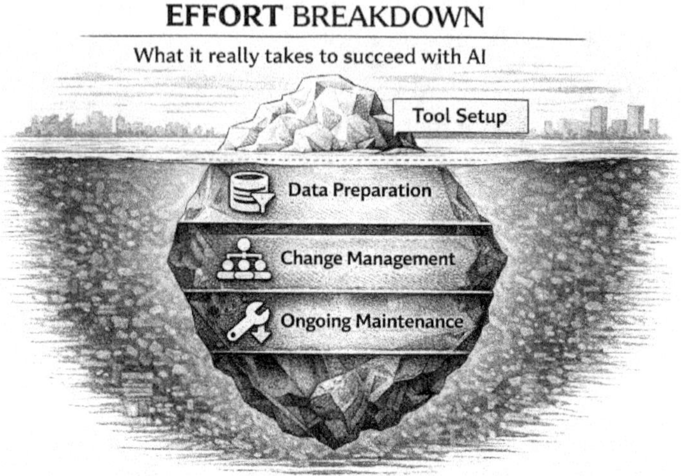

Dimension 3: Risk

Risk captures what could go wrong and how bad it would be. This is often the dimension executives underweight. They get excited about impact and underestimate downside scenarios.

Error Severity: What happens when the AI makes a mistake? An internal document draft with errors is inconvenient. A customer-facing email with errors damages relationships. A financial calculation error could be catastrophic.

Compliance Exposure: Does this touch regulated areas? Healthcare (HIPAA), finance (SOX, GDPR), legal (privilege), and HR (discrimination risk) all require extra caution. Higher compliance stakes mean higher risk.

Data Sensitivity: What data flows through this AI? Public information is low risk. Customer PII, financial data, or trade secrets are high risk. Consider both what the AI sees and what it might inadvertently expose.

Reversibility: Can you undo it if things go wrong? A draft that gets reviewed before sending is highly reversible. An automated response to customers is less so. An AI decision that affects hiring or lending may be legally difficult to reverse.

Risk Scoring Guide

Score	Criteria
1 (Low)	Internal use only, human review before action, no sensitive data, easily reversible
2	Some external exposure but with oversight, minor data sensitivity
3	Customer-facing with limited oversight, some compliance considerations
4	Direct customer impact, regulated data, errors have real consequences
5 (High)	High-stakes decisions, significant compliance risk, sensitive data, hard to reverse

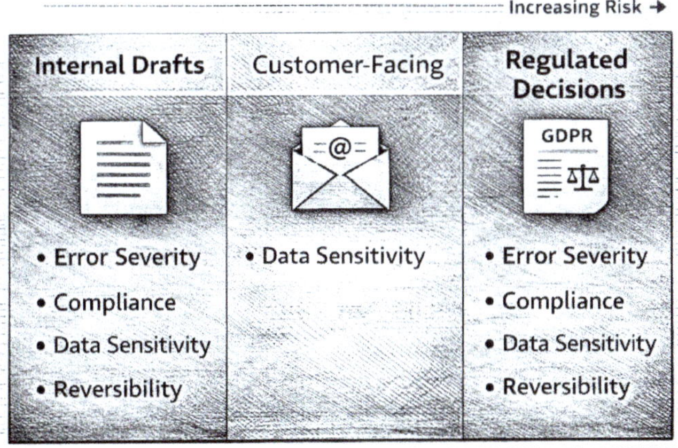

Putting It Together: The Priority Score

Once you've scored each dimension, calculate a simple Priority Score:

Priority Score = Impact × (6 - Effort) × (6 - Risk)

This formula rewards high impact while penalizing high effort and high risk. A project with Impact=5, Effort=2, Risk=1 scores $5 \times 4 \times 5 = 100$. A project with Impact=4, Effort=4, Risk=3 scores $4 \times 2 \times 3 = 24$. The first is clearly a better starting point.

Don't get too attached to the exact numbers. The value is in the structured comparison, not the precision. If two opportunities score within 20% of each other, they're essentially tied, and you should use judgment to pick between them.

The sweet spot for first AI projects: Impact of 3-5, Effort of 1-2, Risk of 1-2. These are opportunities with meaningful value, low implementation friction, and minimal downside. They build confidence and capability for tackling harder projects later.

AI INITIATIVE RESPONSE ACTIONS

ACT	DEFER	PLAN	PAUSE
>35	25–35	15–25	<15
More Value – Less Trouble	Moderate Case for Impact	High Effort or High Risk	Low Impact– High Effort
• Develop the AI solution.	• Monitor the landscape.	• Mitigate risks and difficulty.	• Avoid wasting resources.
• Implement quickly.	• Build a business case.	• Establish clear path.	• Revisit later if needed.

$$\text{SCORE} = \text{IMPACT} \times (6 - \text{EFFORT}) \times (6 - \text{RISK})$$

The 5 Processes Every SMB Should Evaluate First

Based on patterns across hundreds of SMB AI implementations, these five process categories consistently offer the best combination of impact, achievability, and low risk. Not every business will find gold in every category, but most will find at least two or three strong candidates here.

1. Customer Communication Drafting

What it includes: Email responses, proposal drafts, follow-up messages, customer updates, complaint responses, standard correspondence.

Why it's a top candidate: High volume (most businesses send dozens to hundreds of customer communications daily), significant time per message (even "quick" emails take 5-15 minutes to compose thoughtfully), and AI is genuinely good at this with human review.

Typical impact: 50-70% reduction in drafting time, improved consistency in tone and coverage of key points, faster response times.

Risk profile: Low if human reviews before sending. Medium if automating responses without review.

Best starting point within this category: Internal drafts that get reviewed (proposals, important emails to key accounts) rather than high-volume automated responses.

2. Document Summarization and Analysis

What it includes: Summarizing long reports, extracting key terms from contracts, analyzing survey responses, synthesizing research, reviewing meeting transcripts.

Why it's a top candidate: This is AI's wheelhouse. Modern language models can process documents at superhuman speed while maintaining good comprehension. Most professionals spend significant time reading things they don't need to read in full.

Typical impact: 3-5x faster processing of long documents, ability to cover more material, reduced risk of missing key details in dense text.

Risk profile: Low for general summarization. Higher for legal or compliance documents where nuance matters.

Best starting point within this category: Meeting notes and internal reports rather than contracts or regulatory documents.

3. Meeting Support (Prep, Notes, Follow-ups)

What it includes: Pre-meeting research and briefings, generating agendas, taking and organizing meeting notes, drafting follow-up emails and action item summaries.

Why it's a top candidate: Meetings consume enormous executive time, and the administrative work around meetings (prep, notes, follow-up) often gets shortchanged. AI can handle much of this overhead, making meetings more productive.

Typical impact: Better-prepared meetings, more complete notes, faster follow-up. Hard to quantify precisely but executives typically report recovering 3-5 hours weekly.

Risk profile: Low. This is almost entirely internal work with human oversight built in.

Best starting point within this category: Meeting prep briefs (research on attendees, past interactions, likely topics) before moving to note-taking.

4. Research and Information Gathering

What it includes: Market research, competitive analysis, vendor evaluation, regulatory lookups, general business research, synthesizing information from multiple sources.

Why it's a top candidate: Research is time-intensive and often deprioritized. AI can rapidly synthesize information, generate questions you might not have thought to ask, and provide first-pass analysis that accelerates human decision-making.

Typical impact: 70-80% reduction in initial research time. Decisions made with better information because the bar to "do the research" is lower.

Risk profile: Medium. AI can hallucinate facts or miss nuances. All AI research should be verified before acting on it.

Best starting point within this category: Structured research tasks with verifiable outputs (competitive feature comparisons, market sizing) rather than open-ended analysis.

5. Content Creation (Marketing, Internal Comms)

What it includes: Marketing copy, social media posts, blog content, internal announcements, training materials, job descriptions, newsletter content.

Why it's a top candidate: Content creation is a perpetual bottleneck at most companies. There's always more content needed than capacity to create it. AI significantly lowers the production cost of competent first drafts.

Typical impact: 2-3x content output with same team, or significant time savings maintaining current output. Quality varies, though AI content typically needs more editing for voice and specificity.

Risk profile: Low to medium. Brand voice drift is the main risk. Factual claims in content should be verified.

Best starting point within this category: Internal communications and routine content (job postings, standard announcements) before customer-facing marketing.

Process Mining Lite: Finding Hidden Opportunities

The five categories above cover the most common AI opportunities, but your business has unique processes that might be even better candidates. Here are four techniques to surface opportunities you might be missing.

The Time Tracking Method

For one week, ask key team members to track how they spend their time in 30-minute blocks. Don't use fancy software, a simple spreadsheet works. Categorize activities into:

Creating (writing, designing, building),

Processing (reviewing, approving, organizing),

Communicating (meetings, emails, calls), and

Administrating (scheduling, filing, data entry).

Look for any category consuming more than 20% of someone's time. Those are your AI candidates. Especially look for Processing and Administrating: these are often invisible time sinks that AI handles well.

The Frustration Audit

Ask your team: "What task do you dread most? What makes you think 'there has to be a better way'? What would you eliminate if you could?" The emotional signal is valuable. Frustration often indicates tasks that are tedious, repetitive, or cognitively draining in ways that AI might help.

Common frustration responses that signal AI opportunity: "I spend hours every week on..." "I can never catch up on..." "We keep making the same mistakes with..." "Nobody has time to properly..." These are process pain points waiting for a solution.

The Pattern Finder

Look for tasks that have these characteristics. All of them suggest AI suitability:

- **High volume:** Done many times per day/week rather than occasionally

- **Relative consistency:** Follows a recognizable pattern even if details vary

- **Language-heavy:** Involves reading, writing, summarizing, or analyzing text

- **Quality variance acceptable:** 80% quality consistently is better than 100% quality occasionally

- **Human review feasible:** Someone can check the output before it goes external

The Backlog Check

What work is perpetually behind? Most businesses have backlogs that never get addressed: documentation that needs writing, processes that need documenting, emails that need answering, content that needs creating. These backlogs exist because the work is important-but-not-urgent and keeps getting deprioritized.

AI can often tackle backlog items at a fraction of the human time investment. The ROI calculation changes significantly when you're comparing "AI + human review" against "never getting done." That procedure manual no one has time to write? AI can draft it in an afternoon.

Quick Wins vs. Strategic Plays: Balancing Your Portfolio

Not all AI opportunities serve the same purpose. Understanding the difference between quick wins and strategic plays helps you build a balanced AI portfolio.

Quick Wins

Definition: AI applications that can be implemented in days to weeks, deliver immediate value, require minimal change management, and have low risk.

Purpose: Build confidence, demonstrate value, develop organizational AI literacy, create momentum for larger initiatives.

Examples: Using ChatGPT/Claude for email drafting, meeting prep, document summarization. Implementing an AI writing assistant for marketing. Using AI for first-draft job descriptions or internal announcements.

Typical characteristics: Impact scores 2-4, Effort scores 1-2, Risk scores 1-2. Priority Score: 50-100.

Strategic Plays

Definition: AI applications that require significant investment (time, money, organizational change) but deliver transformative value or competitive advantage.

Purpose: Create lasting differentiation, enable new capabilities or business models, achieve step-change improvements in operations or customer experience.

Examples: Custom AI models trained on proprietary data, AI-powered product features, fully automated customer service, predictive systems integrated with operations.

Typical characteristics: Impact scores 4-5, Effort scores 3-5, Risk scores 2-4. Priority Score: varies widely.

Recommended Portfolio Mix by Stage

Stage	Timeline	Quick Wins	Strategic
Starting Out	First 90 days	100%	0%
Building	Months 3-6	70%	30%
Maturing	Months 6-12	50%	50%
Advanced	Year 2+	30%	70%

Quick Wins vs Strategic Plays Matrix

Key Takeaways

Let's distill this module into actionable principles:

1. **Use a framework, not gut instinct.** The Impact/Effort/Risk scoring forces structured thinking and makes it easier to compare opportunities objectively and explain decisions to stakeholders.

2. **Start with the "boring five."** Customer communication, document processing, meeting support, research, and content creation are proven high-value, low-risk starting points for most businesses.

3. **Mine your processes for hidden opportunities.** Time tracking, frustration audits, pattern finding, and backlog checks surface AI candidates specific to your business.

4. **Balance quick wins with strategic plays.** Quick wins build confidence and capability; strategic plays create lasting advantage. You need both, but start with quick wins.

5. **Industry context matters.** The best AI opportunities vary by industry. Use the industry-specific guidance as a starting point, then adapt to your situation.

6. **Your first project sets the tone.** Choose something with high probability of success. Early wins create advocates; early failures create skeptics.

Your 48-Hour Challenge

Before moving to Module 3, complete at least one of these exercises:

1. **Score three opportunities:** Pick three potential AI applications in your business and score each using the Impact/Effort/Risk framework. Which one scores highest? Does that match your intuition?

2. **Run a frustration audit:** Ask your team the frustration questions: What do you dread? What makes you think "there has to be a better way"? Collect and review the responses for AI opportunity signals.

3. **Map the boring five:** For each of the five process categories, identify whether it's a strong, moderate, or weak opportunity in your specific business. Document your reasoning.

What's Next

You now have a prioritized list of AI opportunities. The temptation at this point is to keep analyzing, keep scoring, keep comparing. Don't.

Good enough prioritization beats perfect prioritization that never turns into action.

Module 3 is a week-by-week playbook for your first 90 days of implementation. It's the bridge between "we should do something with AI" and actually doing it.

THE AI OWNER'S MANUAL

The First 90 Days

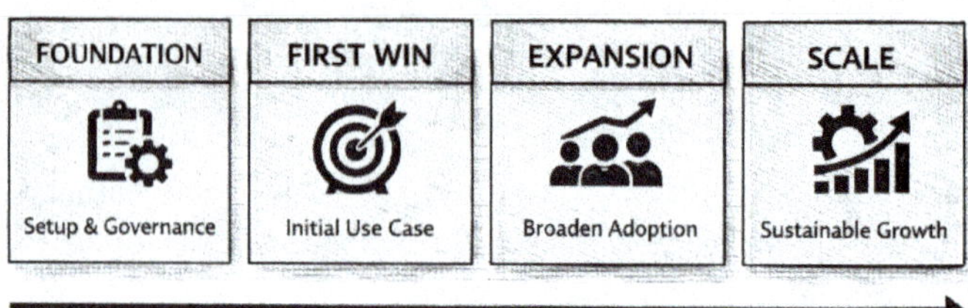

A week-by-week implementation roadmap that takes you from zero to measurable results

Module Overview

You've cut through the hype. You've identified your first AI opportunity. Now comes the question every executive asks: "What exactly do I do on Monday morning?"

This module answers that question with a detailed, week-by-week roadmap for your first 90 days of AI implementation. It's not a theoretical framework: it's a practical playbook built from patterns we have observed across dozens of successful SMB AI adoptions. You'll know exactly what to do in week one, what to expect by week four, and how to measure success by day 90.

The 90-day timeframe is deliberate. It's long enough to achieve meaningful results and build organizational muscle, but short enough to maintain focus and urgency. Most importantly, it's structured to create early wins that build momentum for everything that follows.

Learning Objectives

- √ Execute a structured 90-day AI implementation plan with clear milestones
- √ Select appropriate tools without analysis paralysis using a decision framework
- √ Build internal AI champions who drive adoption beyond the initial pilot
- √ Establish baseline metrics and track progress toward measurable outcomes
- √ Avoid the common pitfalls that derail most AI initiatives

The 90-Day Framework

Before diving into the week-by-week details, let's establish the strategic logic of the 90-day journey. Each phase builds on the previous one, and the sequence matters.

Phase	Timeline	Focus	Key Outcome
Foundation	Weeks 1-2	Setup, governance, baselines	Ready to implement
First Win	Weeks 3-4	Execute first use case	Measurable results
Expansion	Weeks 5-8	Iterate, refine, add users	Proven value, broader buy-in
Scale	Weeks 9-12	Systematize, plan next	Sustainable adoption

The phases are deliberately front-loaded with preparation. Executives often want to skip straight to implementation, but the foundation work in weeks 1-2 prevents the most common failure modes. Think of it like construction: the foundation isn't visible in the final building, but everything depends on it.

Phase 1: Foundation (Weeks 1-2)

The foundation phase establishes everything you need before the first AI tool touches real work. Skip this phase, and you'll spend weeks 3-4 fighting fires instead of generating value.

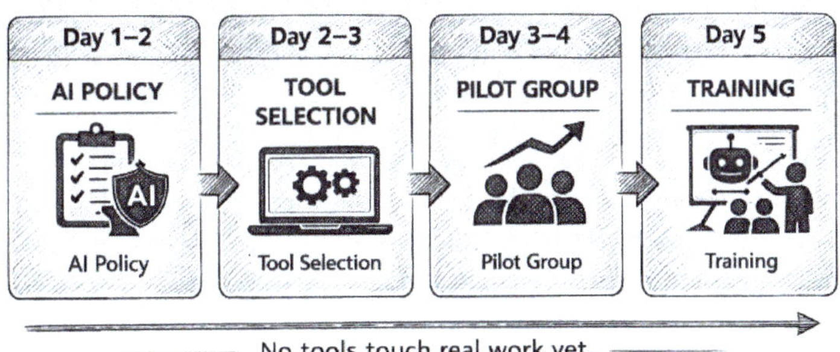

Week 1: Governance and Setup

Week 1 is about creating the environment for safe, effective AI use. By the end of this week, you should have the guardrails in place and the tools ready to go.

Day 1-2: Finalize Your AI Policy

If you haven't already created an AI Acceptable Use Policy, now is the time. Use the template from Module 6 to create a document that addresses data classification, approved tools, prohibited uses, and human review requirements.

Why this matters: Without clear guidelines, employees will either avoid AI entirely (missing the opportunity) or use it recklessly (creating risk). The policy removes ambiguity and gives everyone permission to engage with AI productively.

Day 2-3: Select and Provision Tools

Based on your Module 2 prioritization, you know what use case you're tackling first. Now select the specific tool(s) you'll use. For most SMBs starting with the "boring five" use cases, this means choosing among the major AI assistants.

Tool	Best For	Cost (Dec 2025)
ChatGPT Plus	General drafting, research, coding assistance	$20/month per user
Claude Pro	Long documents, analysis, nuanced writing	$20/month per user
Gemini Advanced	Google Workspace integration, multimodal	$20/month per user
Microsoft Copilot	Microsoft 365 integration, enterprise security	$30/month per user (M365 required)

Recommendation: Start with whatever fits your existing ecosystem. If you're a Google shop, start with Gemini. If you're on Microsoft 365, Copilot has the smoothest integration. If you're ecosystem-agnostic, ChatGPT or Claude are both excellent starting points. The differences between tools matter less than actually using one consistently.

Day 3-4: Identify Your Pilot Group

Don't roll out to the entire company on day one. Identify a small pilot group of 3-5 people who will be the first users. Ideal pilot participants share several characteristics:

- Open to new tools and approaches (not necessarily tech-savvy, just willing)

- Perform the task you're targeting frequently (so they'll get enough practice)

- Willing to provide honest feedback (not just yes-people)

- Respected by peers (their success will influence others)

Among the pilot group, designate one person as your AI Champion. The internal owner who will coordinate feedback, troubleshoot issues, and eventually help train others. This person doesn't need to be technical, but they need to be organized and influential.

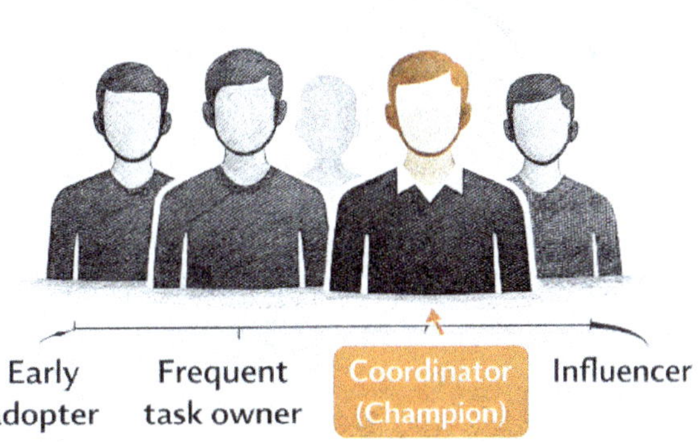

Day 5: Conduct Initial Training

Before the pilot group touches the tool, give them a 60-90 minute training session covering the AI policy (what's allowed and what's not), basic tool orientation (accounts, interface, how to ask questions), the specific use case you're targeting, and prompting basics from Module 5.

Keep training practical. Don't lecture about AI theory: get people using the tool within the first 20 minutes. Have them work through realistic examples from their actual job. The goal is confidence, not mastery.

Week 2: Baselines and Final Prep

Week 2 is about measurement infrastructure and making sure everything is ready for the real work to begin.

Establish Your Baseline Metrics

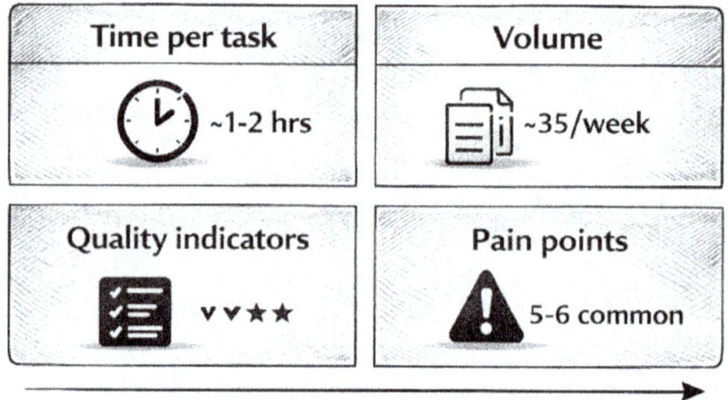

You can't prove ROI without knowing where you started. Before the pilot group starts using AI for real work, capture baseline measurements for your target use case:

1. **Time spent:** How long does the current task take? Have pilot members track time for 3-5 instances of the task this week.

2. **Volume:** How many instances of this task occur per week/month? Check your data.

3. **Quality indicators:** What does "good" look like for this task? Error rates, revision cycles, customer satisfaction scores, whatever is relevant.

4. **Current pain points:** Document the frustrations. These become your success criteria.

Be honest about your baseline. If you're currently averaging 45 minutes per customer email, record that. If quality is inconsistent, note the variance. Accurate baselines make your eventual success credible, inflated baselines make everything look like a failure.

Phase 2: First Win (Weeks 3-4)

This is where the rubber meets the road. The pilot group starts using AI for real work, you collect data, and you iterate based on what you learn. The goal isn't perfection, it's proving value and learning fast.

Week 3: Active Implementation

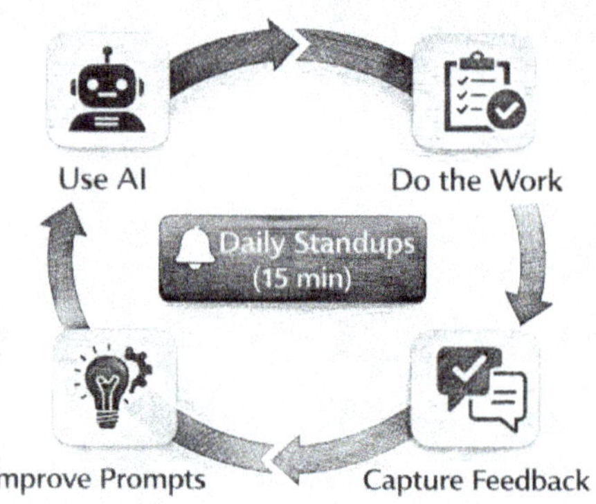

On day one of week 3, the pilot group begins using AI for their target task. Set clear expectations: they should use AI for every instance of the target task during the pilot period. No cherry-picking easy cases or reverting to old methods when things get hard.

The AI Champion should check in with pilot members daily for the first few days. Common early issues include prompt formatting problems (easily fixed with coaching), uncertainty about data classification (refer to policy), and simple frustration when the AI doesn't immediately understand the task (normal prompting is a skill that improves with practice).

Daily Stand-ups (15 minutes)

For the first week of active implementation, hold brief daily check-ins with the pilot group. These shouldn't be formal meetings, just a quick huddle or Slack thread works fine. Cover three questions:

- What worked well yesterday?

- What was frustrating or didn't work?

- Any questions or support needed?

These stand-ups surface problems quickly. If multiple people hit the same issue, you can address it once for everyone. If someone discovers a great prompt or technique, share it immediately.

Week 4: Measure and Iterate

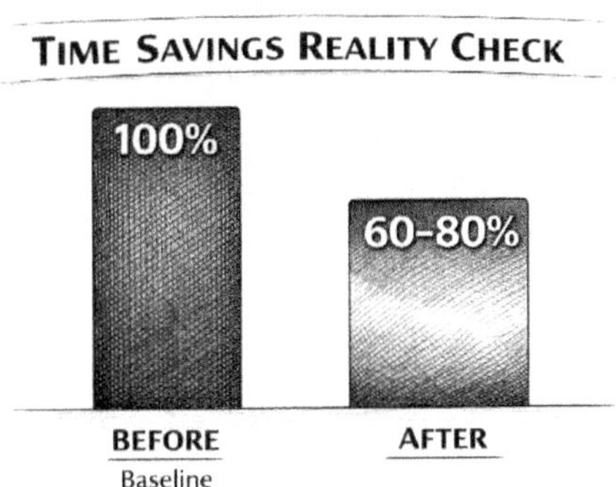

20–40% time savings is a strong first win.

Week 4 shifts focus from implementation to assessment. You have two weeks of data now and that's enough to see patterns and calculate initial results.

Pull together the tracking data from weeks 3-4. Calculate time savings by comparing AI-assisted task time to your baseline. Look at quality by reviewing outputs, are they meeting standards? Better or worse than before? Gather subjective feedback by asking pilot members for honest assessment of the experience.

Don't expect miracles. A realistic first month might show 20-40% time savings with roughly equivalent quality. That's a success. Remember, users are still learning. Results typically improve significantly in months 2-3 as proficiency increases.

Phase 3: Expansion (Weeks 5-8)

Phase 3 is about broadening impact while maintaining quality. You've proven the concept with a small group; now you extend to more users and potentially more use cases.

Double or triple your user base, but don't go company-wide yet. Add users in cohorts. Perhaps one new team or department at a time. This controlled expansion lets you manage training load and catch issues before they affect everyone.

For new users, use your pilot group. Have experienced users co-train or mentor newcomers. This peer-to-peer learning is often more effective than top-down training, and it develops your internal champions.

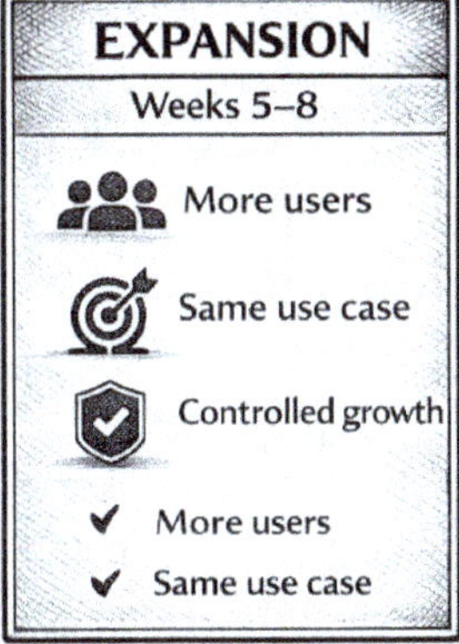

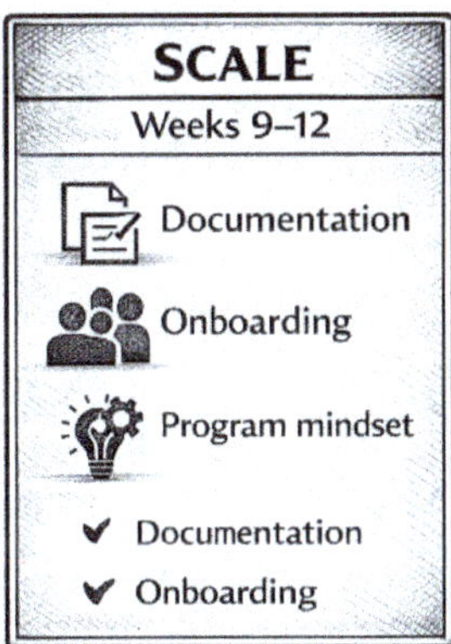

Phase 4: Scale (Weeks 9-12)

The final phase transitions from project to program. You're not just running a pilot anymore. You're building sustainable AI adoption into how your organization works.

Convert your pilot learnings into permanent resources: finalize the prompt library with categories, examples, and usage notes; create an onboarding guide for new users (can be self-serve); write up case studies from successful users (internal marketing); and update the AI policy based on everything you've learned.

This documentation ensures knowledge isn't lost when people change roles or leave. It also makes future expansions much easier. Think of it as a gift to your future self.

Common Pitfalls and How to Avoid Them

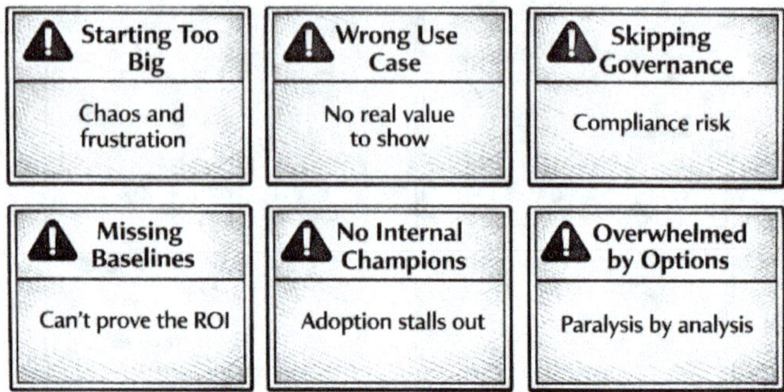

These are the most common failure modes we see in SMB AI implementations. Knowing them in advance helps you avoid them.

Pitfall 1: Skipping the Foundation Phase

What happens: Eager to show results, executives skip governance and training, jump straight to tool deployment. Within weeks, there's confusion about what's allowed, data security incidents, and inconsistent adoption.

How to avoid it: Treat weeks 1-2 as non-negotiable. The foundation work is invisible but essential.

Pitfall 2: Boiling the Ocean

What happens: Instead of focusing on one use case, the organization tries to implement AI everywhere simultaneously. Resources are spread thin, nothing gets done well, and the initiative collapses.

How to avoid it: Force prioritization. One use case for the first 90 days. One pilot group. Success in a narrow scope beats failure in a broad one.

Pitfall 3: No Baseline Metrics

What happens: The pilot runs, people say it "feels faster" and "seems helpful," but there's no data to prove it. When leadership asks for ROI, there's nothing concrete to show.

How to avoid it: Establish baselines in week 2, before implementation begins. Track consistently throughout. Even imperfect data is better than no data.

Pitfall 4: Shadow AI

What happens: While the official pilot proceeds slowly, employees start using their own AI tools without governance. Sensitive data gets pasted into free tools. The official initiative is undermined.

How to avoid it: Acknowledge that people are curious and want to use AI now. The governance framework should enable appropriate use, not just prohibit risky use. Move fast enough that the official channels meet employee demand.

Pitfall 5: Training Once and Forgetting

What happens: Initial training happens in week 1, then nothing. Users don't develop proficiency. Prompts stay basic. The tool gets used at 20% of its potential, or even worse, stops being used at all.

How to avoid it: Training is ongoing, not one-time. The daily standups in week 3 are mini-training sessions. The prompt library is continuous training. Proficiency develops over months, not days.

Pitfall 6: Ignoring Change Management

What happens: The technology works fine, but people don't adopt it. Some resist openly; others nod along but quietly continue old methods. Adoption plateaus at 30% of the target group.

How to avoid it: Treat this as a change initiative, not just a technology deployment. Address the "what's in it for me" question explicitly. Involve respected peers as champions. Celebrate early wins publicly. Module 8 covers change management in depth.

Key Takeaways

1. **Foundation before implementation.** Weeks 1-2 establish governance, select tools, train users, and capture baselines. This work prevents the problems that derail most AI initiatives.

2. **Start narrow, expand systematically.** One use case, one pilot group, one tool. Prove value in a controlled environment before broadening scope.

3. **Measure from day one.** Baselines before implementation, tracking during, ROI calculation after. Data makes the case for expansion and continued investment.

4. **Iterate constantly.** Daily standups in early weeks, weekly reviews later, retrospectives at phase boundaries. AI implementation is a learning process, not a one-time deployment.

5. **Build institutional knowledge.** Prompt libraries, training materials, documented workflows. Knowledge that stays in people's heads leaves when they do.

6. **This is change management, not just technology.** Address resistance, celebrate wins, develop champions. The human factors determine success more than the technology.

Your 48-Hour Challenge

Before moving to Module 4, complete at least one of these exercises:

1. **Draft your AI policy.** Use the template from Module 6 to create a first draft of your AI Acceptable Use Policy. It doesn't need to be perfect. Get something on paper that you can refine.

2. **Identify your pilot group.** List 3-5 people who would make good pilot participants for your first AI initiative. Note why each person is a good fit. Identify your potential AI Champion.

3. **Calculate your baseline.** For your target use case (from Module 2), estimate the current time investment. How many people do this task? How often? How long does it take? What's the total weekly/monthly time investment?

What's Next

You've got the 90-day plan. But here's a question that comes up in almost every implementation I've been part of: what is *my* role in all this? Not the organization's role. Mine. As CFO, or COO, or head of marketing.

That's exactly what Module 4 answers. It has role-specific playbooks with actual prompts you can start using today, whether you're running finance, operations, sales, HR, or IT.

THE AI OWNER'S MANUAL

AI for Your Leadership Team

Role-specific playbooks so every executive knows exactly how AI applies to their function

Module Overview

You've now learned what AI can do (Module 1), identified your best opportunities (Module 2), and mapped out your first 90 days of implementation (Module 3). But there's a question we haven't directly addressed: What about you? What about your CFO, your COO, your head of sales?

AI isn't just a tool for "the team" to use while leadership watches from above. The executives who will thrive in the AI era are those who personally integrate these tools into their own work. Modeling the behavior, they want to see and discovering opportunities invisible to those who only read about AI.

This module provides role-specific playbooks for each member of your leadership team. Whether you're the CEO setting strategy, the CFO analyzing financials, or the CHRO managing talent, you'll find concrete applications tailored to your function. These aren't generic suggestions. They're specific workflows, prompts, and tools that your peers are using right now to work smarter.

Learning Objectives

- ✓ Understand how AI applies differently to each executive function
- ✓ Identify the highest-value AI applications for your specific role
- ✓ Apply ready-to-use workflows and prompts for common leadership tasks
- ✓ Lead by example in AI adoption across your organization

The AI Leadership Matrix

Before diving into role-specific playbooks, it helps to see the market. Different executive roles have different AI opportunities based on the nature of their work. This matrix maps high-value AI applications across the C-suite:

Role	Communication	Analysis	Knowledge
CEO	Board updates, investor memos, stakeholder comms	Strategic synthesis, competitive intel	Market trends, industry research
CFO	Financial narratives, audit responses	Variance analysis, scenario frameworks	Regulatory updates, benchmarking
COO	Policy docs, operational updates	Process analysis, vendor comparison	Best practices, compliance
CMO	Content creation, sales enablement	Customer feedback, campaign performance	Market research, competitor tracking
CHRO	Job postings, policy updates	Engagement surveys, comp analysis	Employment law, industry standards

Notice the patterns: every role benefits from communication assistance, but the analytical applications vary significantly. The CEO needs strategic synthesis; the CFO needs financial modeling; the COO needs process analysis. The playbooks that follow dive deeper into each role.

The CEO/Owner Playbook

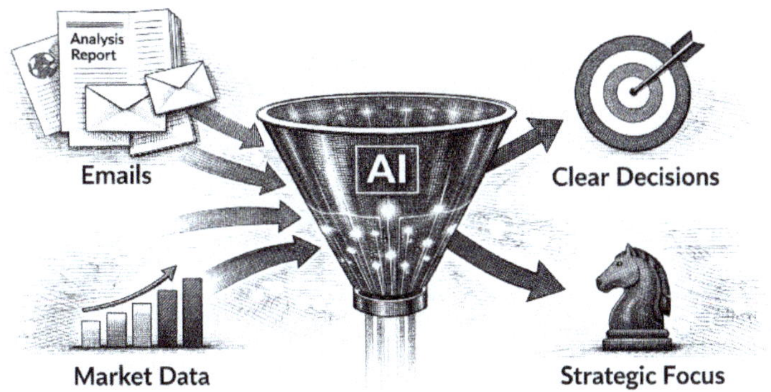

As CEO, your scarcest resource is attention. Every hour spent on tasks that don't require your unique judgment is an hour stolen from strategic thinking, relationship building, and the decisions only you can make. AI's highest value for you isn't in doing new things. It's in compressing the time required for essential activities.

Core Applications

Strategic Synthesis

You receive more information than you can possibly process: board reports, market analyses, competitive intelligence, internal updates. AI excels at synthesizing large volumes of information into actionable intelligence. Feed it a 50-page industry report and ask for the three insights most relevant to your strategic priorities. Paste in multiple analyst reports and request a comparative analysis highlighting where they agree and disagree.

Board and Investor Communication

Board updates, investor memos, and stakeholder communications demand precision and polish. But the first draft is often the hardest part. Use AI to draft board update emails, create presentation outlines, or prepare talking points for difficult conversations. Always review and personalize but start from a solid foundation rather than a blank page.

Decision Preparation

Before major decisions, use AI as a thinking partner. Describe a strategic dilemma and ask for considerations you might be missing. Request a devil's advocate perspective on a planned initiative. Have it generate questions that a skeptical board member might ask. This isn't about having AI make decisions. It's about ensuring you've considered multiple angles.

Sample CEO Prompts

1. *"Summarize this 40-page market analysis. Focus on trends affecting our mid-market segment and any threats we should address in the next 12 months."*

2. *"Draft talking points for a board meeting where I need to explain why we're delaying the product launch by one quarter. Acknowledge the concern, explain the reasoning, and emphasize the long-term benefit."*

3. *"I'm considering acquiring a smaller competitor. What questions should I be asking that I might not have thought of? What due diligence areas are commonly overlooked in similar acquisitions?"*

The CFO/Finance Playbook

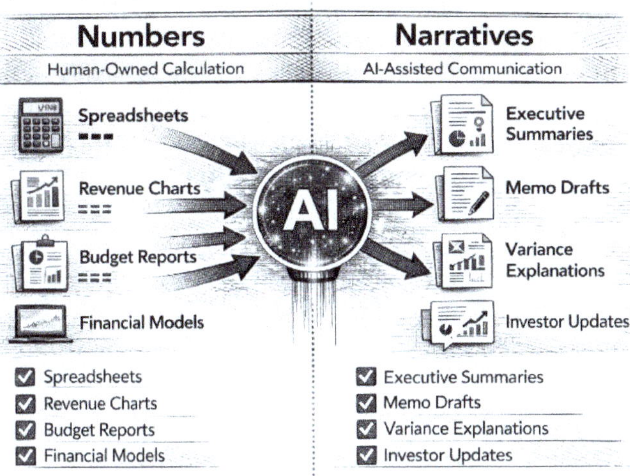

Finance leaders operate in a world of precision where a misplaced decimal can cascade into serious consequences. This makes many CFOs appropriately cautious about AI. The key is using AI where it excels (synthesis, communication, pattern recognition) while keeping humans firmly in control of the numbers themselves.

Core Applications

Financial Narrative

Numbers tell a story, but humans need narratives. AI can help translate financial data into clear explanations for non-financial stakeholders. Paste in quarterly results and ask for a narrative explanation suitable for an all-hands meeting. Request variance analysis summaries that explain the "why" behind the numbers. The AI won't calculate. But it will help communicate what the calculations mean.

Report and Memo Drafting

CFOs produce enormous volumes of written communication: board reports, audit responses, budget justifications, policy memos. AI speed ups the drafting process. Provide the key points you need to convey, specify the audience and tone, and let AI create the first draft. You edit for accuracy and nuance; AI handles structure and polish.

Scenario Analysis Frameworks
AI can help structure scenario analyses and sensitivity frameworks. Not running the actual numbers, but organizing the approach. Describe a business scenario and ask AI to identify the key variables you should model. Request a framework for stress-testing assumptions. Use it to generate the qualitative considerations that should accompany quantitative analysis.

Sample CFO Prompts

1. *"Here are our Q3 results compared to budget. Write a narrative explanation for the board that highlights the key variances and explains our revised forecast. Tone should be confident but realistic."*

2. *"We're implementing a new ERP system. Draft a project justification memo that addresses cost, timeline, risk, and expected ROI. This will go to the executive committee."*

3. *"I need to prepare for a conversation with auditors about our revenue recognition approach. What questions are they likely to ask, and what documentation should I have ready?"*

The COO/Operations Playbook

Operations leaders live in the details: the processes, the workflows, the thousand small things that need to go right for the business to function. This gives you an advantage in AI adoption: you see exactly where time is wasted and can identify specific, measurable opportunities for improvement.

Core Applications

Process Documentation

Every operations leader knows the pain of undocumented processes, the tribal knowledge that exists only in people's heads. AI can help capture and formalize this knowledge. Describe a process verbally (or paste rough notes) and ask AI to create a formal SOP. Have it generate process flowcharts, checklists, or training materials from your existing documentation.

Vendor and Contract Analysis

COOs often manage dozens of vendor relationships. AI can accelerate contract review, compare vendor proposals, and identify key terms across multiple agreements. Paste a vendor contract and ask for a summary of key obligations, renewal terms, and potential risks. Compare two competing proposals side by side.

Operational Communication
From policy updates to incident reports to change management communications, operations generates substantial written output. Use AI to draft policy documents, create announcement templates, or summarize operational issues for executive consumption. Your job is to ensure accuracy and appropriateness; AI handles the initial construction.

Sample COO Prompts

1. *"Here are my notes from shadowing our shipping team. Turn these into a formal standard operating procedure with numbered steps, quality checkpoints, and exception handling."*

2. *"Compare these two vendor proposals for our logistics software. Create a table showing pricing, features, implementation timeline, and contract terms. Note any significant differences."*

3. *"Draft a company-wide announcement about our new expense reporting policy. The key changes are [list changes]. Tone should be clear and firm but not punitive."*

The CMO/Sales Playbook

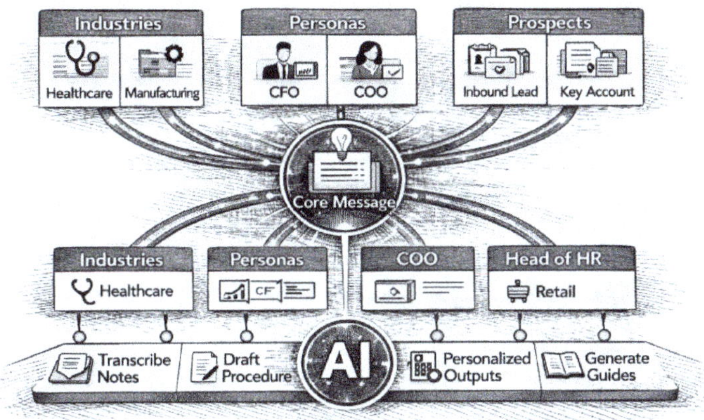

Marketing and sales leaders were early AI adopters, content creation is an obvious use case. But many have only scratched the surface, using AI for basic copywriting while missing higher-value applications in strategy, analysis, and personalization.

Core Applications

Content Strategy and Creation

Beyond basic copywriting, AI can help develop content strategies, create editorial calendars, and generate content briefs for your team. Describe your target audience and business objectives, then ask for a content plan. Use AI to repurpose content across formats. Turning a whitepaper into a blog series, or a webinar into social posts.

Customer Intelligence

AI can synthesize customer feedback, reviews, and support tickets to surface patterns you might miss. Paste customer feedback from multiple sources and ask for themes, sentiment analysis, or specific improvement suggestions. Use it to prepare for customer calls by summarizing the relationship history and identifying potential opportunities or concerns.

Sales Enablement

Create personalized sales materials at scale. Feed AI a prospect's website and ask for a customized value proposition. Generate objection responses, proposal drafts, or follow-up sequences. The key is personalization. Moving beyond generic templates to materials that show you understand the specific prospect's situation.

Sample CMO/Sales Prompts

1. *"Here's our company overview and target customer profile. Create a 3-month content calendar with blog topics, social posts, and email campaigns aligned with our sales cycle."*

2. *"I have a meeting with [Prospect Company] tomorrow. Here's their website and LinkedIn page. Identify their likely pain points and draft 3 questions that would demonstrate understanding of their business."*

3. *"Analyze these 50 customer support tickets. What are the top 5 product complaints? Summarize each with frequency and suggested improvements."*

The CHRO/HR Playbook

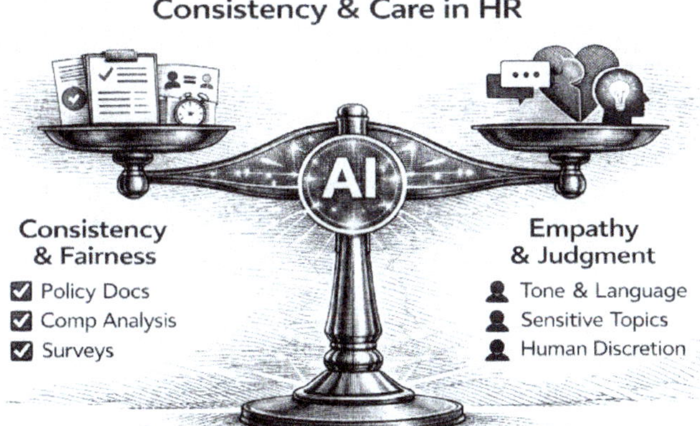

Human Resources might seem like an unexpected place for AI, given its focus on, well, humans. But HR leaders often face crushing administrative workloads that distract from strategic people initiatives. AI can help reclaim that time while also improving consistency and reducing bias in certain processes.

Core Applications

Job Descriptions and Postings

Writing job descriptions is time-consuming and often produces inconsistent results. AI can generate first drafts based on role requirements, ensure consistent formatting across positions, and flag potentially biased language. Provide the key requirements and let AI create the structure; you refine for culture fit and accuracy.

Policy Documentation

Employee handbooks, policy updates, and HR communications require careful language. AI can help draft policies from outlines, update existing policies for new situations, and ensure consistent language across HR documents. Always have legal review policies before publication, but AI can significantly reduce the drafting time.

Employee Communication

HR sends more company-wide communications than any other function. Use AI to draft announcements, create FAQ documents for policy changes, or generate talking points for managers delivering difficult messages. AI helps ensure completeness and appropriate tone while you focus on the human elements.

Sample CHRO Prompts

1. *"Create a job description for a Senior Product Manager. Key requirements are [list requirements]. The tone should reflect our collaborative culture. Flag any language that might inadvertently discourage diverse candidates."*

2. *"We're changing our PTO policy from accrued to unlimited. Draft an announcement email that explains the change, addresses likely concerns, and includes an FAQ section."*

3. *"Create talking points for managers who need to deliver performance improvement plans. Include how to set the right tone, key points to cover, and responses to common employee reactions."*

The CTO/IT Playbook

Technology leaders often have more hands-on AI experience than other executives, but that same familiarity can lead to blind spots. You know AI can write code; but are you using it for the documentation, communication, and strategic thinking that consume your time?

Core Applications

Technical Documentation

Documentation is every tech team's perpetual backlog item. AI can help clear that backlog. Generate API documentation from code comments, create architecture decision records from meeting notes, or produce onboarding guides for new team members. The documentation that never gets written can now get written. With AI doing the heavy lifting.

Code Review Assistance

AI coding assistants have become essential for development teams. But they're also useful for CTOs who need to understand code without writing it. Use AI to explain what a piece of code does, identify potential issues, or suggest improvements. Stay connected to technical details without needing to spend hours reading code yourself.

Technology Strategy Communication

CTOs must translate technical complexity for non-technical stakeholders. AI can help bridge this gap. Drafting board updates on technology initiatives, creating executive summaries of technical proposals, or preparing business cases for infrastructure investments. Your expertise provides the substance; AI helps with the translation.

Sample CTO Prompts

1. *"Here's our authentication module code. Create developer documentation that explains the flow, key functions, and integration points. Include code examples."*

2. *"Explain this pull request to me in plain language. What is it trying to accomplish? Are there any potential issues or alternative approaches?"*

3. *"Draft a board presentation slide deck outline for our cloud migration project. Focus on business benefits, risk mitigation, and timeline. Avoid technical jargon, this is for non-technical board members."*

Tool Recommendations by Role

While the major AI assistants (ChatGPT, Claude, Gemini, Copilot) work well for all roles, certain tools offer specific advantages:

Role	Recommended Tool	Why
CEO	Claude Pro or ChatGPT Plus	Strong at synthesizing long documents; nuanced communication
CFO	Microsoft Copilot	Excel integration for financial narratives; enterprise security
COO	Claude Pro or ChatGPT Plus	Excellent at process documentation; handles long contracts
CMO	ChatGPT Plus + nano banana or Imagen	Content creation strength; image generation included
CHRO	Claude Pro	Strong at nuanced HR language; careful with sensitive topics
CTO	Claude Pro + GitHub Copilot	Technical depth; code-aware; excellent documentation

Start with the tool that best fits your existing ecosystem, integration matters more than marginal capability differences. If you're already in Microsoft 365, start with Copilot. If you're Google-based, start with Gemini. Most executives find that any of the major tools serves their needs well once they develop prompting skills.

Key Takeaways

1. **Lead by example.** Your personal AI adoption signals to the organization that this is a priority. The executives who actually use AI are far more effective at driving organizational adoption than those who only mandate it.

2. **Focus on your unique pain points.** The best AI applications vary by role. Use the playbooks above as starting points, then adapt based on where you personally spend time that doesn't require your unique judgment.

3. **AI assists, you decide.** Especially at the executive level, AI should inform decisions, not make them. Use it to ensure you've considered multiple perspectives and synthesized available information, but the judgment remains yours.

4. **Communication is the universal win.** Regardless of role, drafting communications is where most executives see immediate value. Start there if you're unsure where to begin.

5. **Review everything.** The stakes are higher at the executive level. Every AI output should be reviewed before it represents you or your function. This isn't a limitation. It's appropriate oversight.

6. **Share what works.** When you find an effective prompt or workflow, share it with your peers. Executive AI adoption accelerates when leaders learn from each other's experiments.

Your 48-Hour Challenge

Before moving to Module 5, complete at least one of these exercises:

1. **Try your role's prompts**: Pick two prompts from your role's playbook and use them for real work this week. Note what worked and what needed adjustment.
2. **Identify your top time drain**: List the three tasks that consume most of your time each week. For each, brainstorm how AI might assist. Test one.
3. **Share a win**: Use AI to help with a real task, then share the experience with a peer executive. Discuss what worked and what you learned.

What's Next

Here's something I see constantly: an executive tries one of the prompts from the last module, gets a mediocre result, and concludes the tool isn't that useful. Nine times out of ten, the issue isn't the tool. It's how they asked.

Module 5 is about prompting, the skill of getting AI to actually give you what you want. It sounds simple, but the difference between a lazy prompt and a well-structured one is the difference between a generic first draft and something you'd actually send.

THE AI OWNER'S MANUAL

Communicating with AI

The essential skill of prompt engineering, how to ask AI tools the right way to get consistently excellent results

Module Overview

In Modules 1 through 4, you learned what AI can do, where to focus, how to implement it, and how it applies to your specific role. But there's a skill that underlies all of it, one that separates people who get mediocre results from AI from those who get exceptional ones.

That skill is prompt engineering: the art and science of communicating with AI tools effectively.

Don't let the technical-sounding name intimidate you. Prompt engineering isn't coding. It's communication. It's learning how to give clear instructions to a very capable but very literal assistant. The same skills that make you effective at delegating to humans (clarity, specificity, context) make you effective at working with AI. The difference is that AI requires you to be even more explicit about what you want.

This module will transform your AI interactions. You'll learn a universal framework that works across ChatGPT, Claude, Gemini, and Copilot. You'll understand why some prompts produce brilliant results while others produce garbage. And you'll walk away with ready-to-use templates for the business tasks you face every day.

Learning Objectives

- ✓ Understand why prompt quality determines AI output quality
- ✓ Apply a universal prompt structure that works across all major AI platforms
- ✓ Master the five essential principles of effective prompting
- ✓ Use ready-to-customize templates for common business tasks
- ✓ Build an organizational prompt library that captures institutional knowledge

Why Prompting Matters

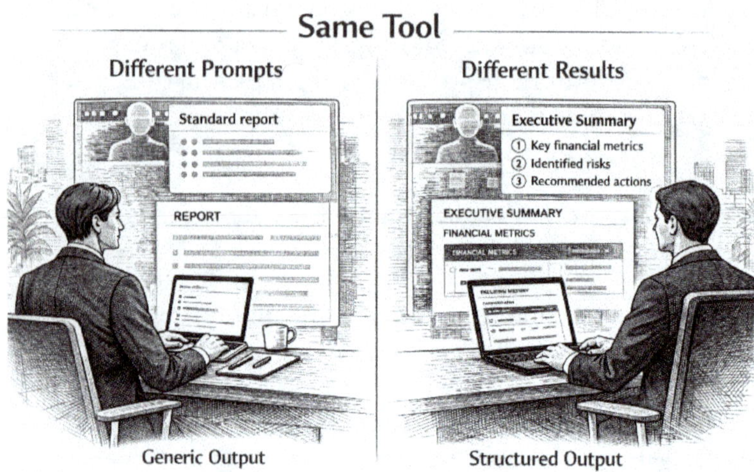

Here's a reality that surprises most executives: two people using the exact same AI tool with the exact same subscription will get wildly different results. The difference isn't the technology, it's how they communicate with it.

Consider this analogy. Imagine you hired a brilliant but brand-new employee who knows nothing about your company, your industry, or your preferences. If you asked them to "write a report," you'd get something generic and probably useless. But if you told them exactly what the report should cover, who would read it, what tone to use, how long it should be, and showed them an example of what good looks like. You'd get something useful.

AI works the same way. It's extraordinarily capable but has no context unless you provide it. A vague prompt produces vague results. A specific, well-structured prompt produces specific, useful results.

The Garbage In, Gold Out Problem

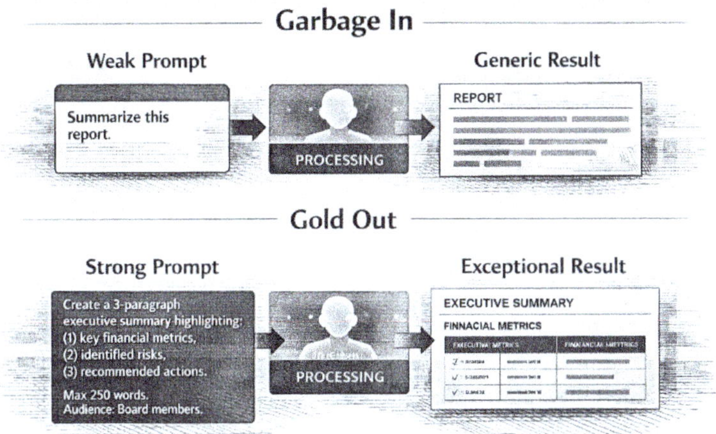

Most people blame the AI when they get poor results. "It didn't understand what I wanted." "The output was too generic." "It missed the point entirely." In almost every case, the problem wasn't the AI; it was the prompt.

Watch the difference:

Weak Prompt	Strong Prompt
"Summarize this document."	*"Create a 3-paragraph executive summary highlighting: (1) key financial metrics, (2) identified risks, (3) recommended actions. Maximum 250 words. Audience: Board members."*
"Help me write an email."	*"Write a professional email to a vendor requesting a 15% discount on our renewal. Tone: firm but collaborative. Length: 150 words. Include specific value we've delivered as their customer."*

Weak Prompt	Strong Prompt
"Analyze this data."	*"Analyze the sales data and identify: top 3 performing regions, month-over-month trends, and anomalies requiring investigation. Present findings in a table with 2-3 key insights."*

The weak prompts leave everything to chance. The strong prompts specify exactly what success looks like. Same AI, significantly different results.

The Universal Prompt Structure

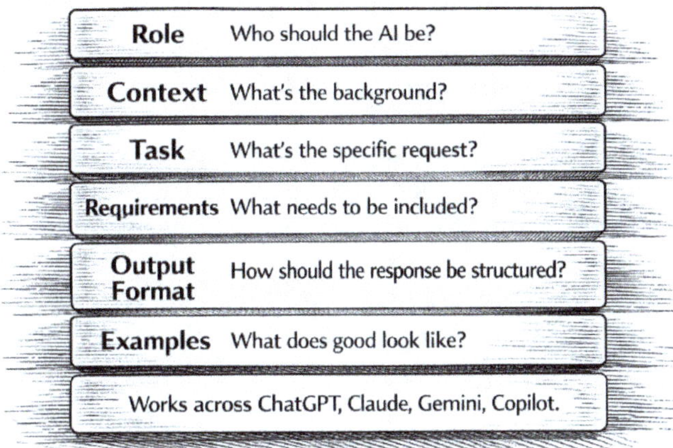

Every effective prompt (whether you're using ChatGPT, Claude, Gemini, Grok, or Copilot) follows the same basic structure. Think of it as a framework you can adapt to any situation. Not every prompt needs every element, but knowing the full structure helps you diagnose why a prompt isn't working.

The Six Elements of an Effective Prompt

1. Role: Who should the AI be? Telling the AI to act as a specific expert focuses its responses. "You are a financial analyst with expertise in SaaS metrics" produces different output than "You are a marketing copywriter."

2. Context: What background does the AI need? Include key facts, relevant constraints, and any information the AI couldn't know without being told. Remember, AI has no access to your company's internal information unless you provide it.

3. Task: What exactly do you want done? State your request clearly in one sentence. This is the core of your prompt, make it unambiguous.

4. Requirements: What should be included or avoided? Specify constraints, must-haves, and things you explicitly don't want. The more specific, the better.

5. Output Format: How should the response be structured? Specify length (word count, paragraphs, pages), format (prose, bullets, table, JSON), sections or headers, and tone (formal, conversational, technical).

6. Examples: What does good look like? When you need consistent format across multiple outputs, include sample input/output pairs. This is the single most effective technique for format adherence.

The Framework in Action

Here's how the framework looks for a real business task, analyzing customer feedback:

[ROLE]
You are a customer insights analyst specializing in B2B software feedback.

[CONTEXT]
We're a project management software company preparing for our quarterly product review. We've collected 200 customer feedback responses. Key priorities this quarter: improving collaboration features and reducing onboarding time.

[TASK]
Analyze the attached customer feedback and identify the top themes, prioritized by frequency and alignment with our quarterly goals.

[REQUIREMENTS]

Include specific customer quotes as evidence. Flag any feedback related to competitor comparisons. Do not include feedback about pricing, that's handled separately.

[OUTPUT FORMAT]
Provide a 500-word analysis organized as: (1) Top 5 Themes table with frequency and example quote, (2) Alignment with Q4 Priorities section, (3) Recommended Actions list with 3-5 specific suggestions.

[EXAMPLE]
Theme: Real-time editing | Frequency: 34 mentions | Quote: "We need to see changes instantly without refreshing." | Priority Alignment: High (collaboration)

Notice how every element serves a purpose. The role focuses the AI's expertise. The context provides necessary background. The task is crystal clear. The requirements prevent common mistakes. The output format ensures usable structure. The example demonstrates exactly what you want.

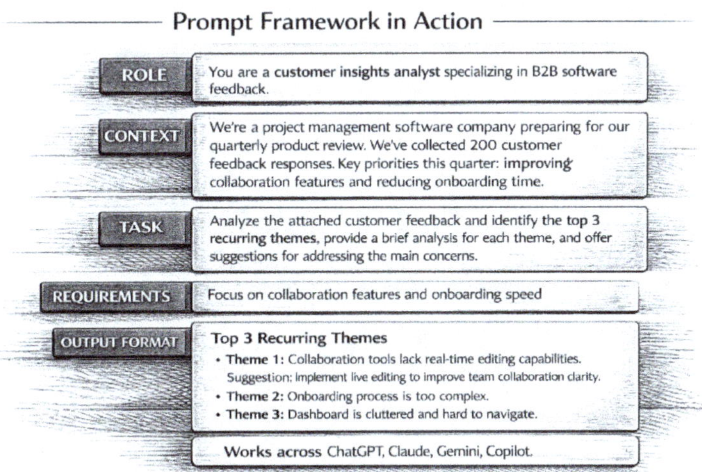

The Five Essential Principles

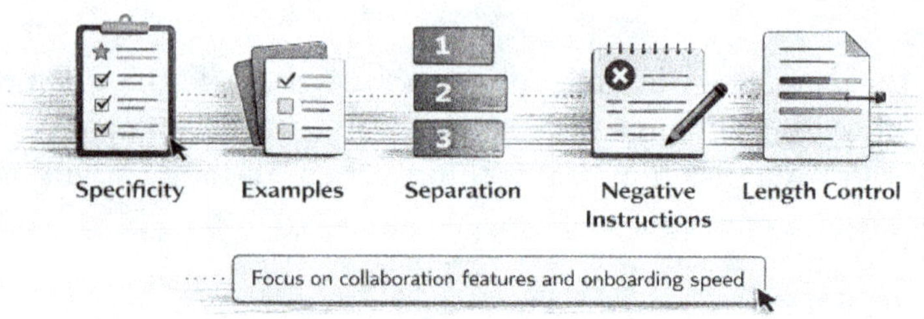

Beyond the framework structure, five principles consistently separate effective prompts from ineffective ones. Master these and you'll significantly improve your AI results.

Principle 1: Be Specific, Not Vague

Vague instructions produce vague results. Every time you use words like "good," "better," "comprehensive," or "detailed" without defining what those mean, you're leaving interpretation to chance.

Instead of "Make it more professional," say "Use formal business language, avoid contractions, and address the reader as 'you' rather than using first person." Instead of "Improve this," specify exactly what improvement means: "Strengthen the opening to immediately state the key finding, reduce word count by 20%, and add a clear call to action."

The test: Could someone else read your prompt and know exactly what you want? If not, add specifics.

Principle 2: Show, Don't Just Tell

Examples are the most powerful prompting technique available. Yet most people skip them. When you show AI what good looks like, it can pattern-match to produce similar results.

This is especially important for format-sensitive tasks. If you need customer feedback categorized a specific way, show one or two examples of correctly categorized feedback. If you need emails in a particular style, include a sample email. If you need data presented in a specific table format, mock up one row of that table.

Two to three examples is usually enough. More than that rarely improves results and just makes your prompt longer.

Principle 3: Separate Instructions from Content

When you're asking AI to process a document, email, or data set, clearly separate your instructions from the content being processed. Use consistent delimiters: dashes, hash marks, or XML-style tags.

INSTRUCTIONS: Review the contract below and identify any liability clauses.

--- CONTRACT TEXT:
[paste contract here]
--- END OF CONTRACT

This separation prevents the AI from confusing your instructions with the content it's supposed to analyze. It's a simple technique that significantly improves accuracy on document processing tasks.

Principle 4: Specify What NOT to Do

Positive instructions tell AI what you want. Negative instructions prevent what you don't want. Both are essential.

AI models have default behaviors that may not match your needs. If you don't want generic advice, say so: "Do not include general suggestions that could apply to any company." If you don't want hedging, say so: "Do not use phrases like 'it depends' without immediately explaining what it depends on." If you don't want marketing language, say so: "Avoid superlatives and promotional tone."

Think about what typically goes wrong with AI outputs in your domain and explicitly prohibit those behaviors.

Principle 5: Control Length Explicitly

Never leave length to chance. AI has no intuition for "appropriate length", it will either over-deliver or under-deliver unless you specify.

When You Need	Say This
Very short	*"Answer in 1-2 sentences."*
Brief	*"Limit response to 100 words."*
Moderate	*"Provide 2-3 paragraphs."*
Detailed	*"Write 500-750 words."*
Structured	*"Create 5 sections, each 100-150 words."*

Common Mistakes and How to Fix Them

Even experienced AI users make these mistakes. Recognizing them helps you diagnose problems quickly.

Mistake 1: Contradictory Instructions

The problem: "Be thorough but keep it brief." "Provide detailed analysis in a few sentences." These contradictions confuse the AI and produce mediocre results that satisfy neither goal.

The fix: Choose one priority. If you need brevity, sacrifice comprehensiveness and specify exactly how brief. If you need comprehensive, accept longer output and specify structure to keep it organized.

Mistake 2: Assuming Context

The problem: Referencing "our project," "the usual format," or "like last time" without providing details. AI has no memory of your previous conversations (unless you're using specific memory features) and knows nothing about your company.

The fix: Provide all necessary context every time. If you find yourself repeatedly typing the same context, create a saved prompt template that includes your standard background information.

Mistake 3: Burying the Task

The problem: Providing paragraphs of context before finally getting to what you want. The AI may lose focus on the actual task or give disproportionate weight to early information.

The fix: Put your task first, context second. Lead with what you want done, then provide the background needed to do it well.

Mistake 4: One-Shot Complex Tasks

The problem: Expecting AI to produce perfect output on the first try for complex tasks. Resulting in frustration when the output isn't quite right.

The fix: Use iterative refinement. Start with an outline, review it, then expand. Or generate a first draft, identify what needs improvement, then specifically request those changes. Complex tasks often benefit from 2-3 rounds of refinement.

Mistake 5: No Output Format

The problem: Leaving structure to chance. The AI might produce prose when you wanted bullets, or a wall of text when you wanted sections.

The fix: Always specify output format. If you don't care about format, you probably still want some constraints. At minimum, specify length.

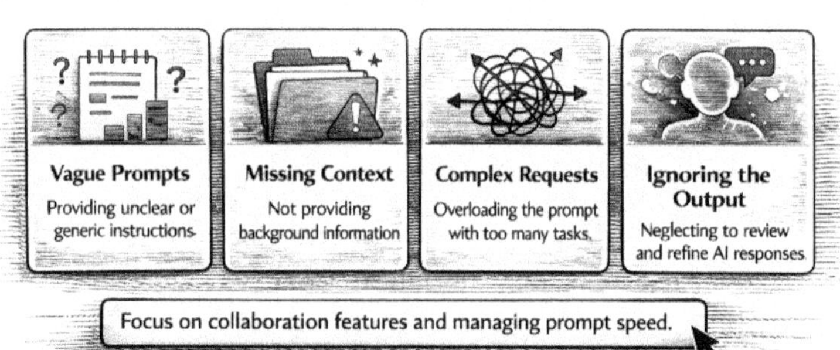

Business Use Case Templates

These templates are ready to use. Copy them, fill in the bracketed sections with your specific information, and adjust as needed for your context.

Document Analysis Template

Review the attached [document type] and provide:

1. SUMMARY (100 words max): Key points a decision-maker needs to know.

2. CRITICAL ITEMS (bullet list): Items requiring action, with page/section references.

3. QUESTIONS/CONCERNS: Ambiguities, missing information, or potential issues.

Quote specific text when referencing the document.

Email Drafting Template

Write an email with these parameters:
- To: [recipient role/relationship]
- Purpose: [specific goal of the email]
- Tone: [formal / friendly / urgent / diplomatic]
- Key points to include: [list main points]
- Length: [short / medium / detailed, or specific word count]
- Call to action: [what you want recipient to do]

Data Analysis Template

Analyze the provided data and deliver:

1. KEY FINDINGS (3-5 bullets): Most important patterns or insights.

2. ANALYSIS: [Specify: trends, comparisons, anomalies, correlations, whatever matters for your use case]

3. RECOMMENDATIONS: Actionable next steps based on the data.

Present numerical findings in a table. Flag any data quality issues that affect conclusions.

Decision Support Template

Help me evaluate: [decision to make]

Context: [relevant background information]

Options being considered: 1) [A] 2) [B] 3) [C]

For each option, analyze:

• Pros and cons

• Risks and mitigation strategies

• Resource requirements

Conclude with a recommendation and rationale. Be direct about which option you'd choose and why.

Meeting Preparation Template

Help me prepare for a meeting about [topic].

Attendees: [who will be there and their roles/interests]

My goal: [what I want to achieve]

Known challenges: [obstacles or concerns to address]

Provide:

1. Key talking points (5 max)

2. Likely objections and responses

3. Questions I should be prepared to answer

Building a Prompt Library

Individual prompting skill is valuable. Organizational prompting capability is transformative.

The companies getting the most value from AI aren't just training individuals. They're building shared libraries of effective prompts. When someone discovers a prompt that works well for a common task, that discovery benefits everyone. When someone invests time perfecting a complex prompt, that investment pays dividends across the organization.

What to Include in Your Library

Task-specific prompts: Prompts for common tasks in your organization. Processing certain document types, generating specific reports, handling particular customer interactions. These should include the full context and requirements that make them work in your environment.

Role-specific prompts: Prompts tailored to specific functions. Your sales team's demo follow-up prompt will differ from your support team's ticket response prompt. Group prompts by role for easy discovery.

Template prompts: Prompts that work for categories of tasks with bracketed sections for customization. The templates in this module are examples, they become even more valuable when adapted with your company's specific context.

How to Build Your Library

Start with high-frequency tasks: Identify the 10-20 tasks your team does repeatedly. Build and refine prompts for those first. A great prompt for a daily task delivers more value than a perfect prompt for something you do quarterly.

Capture what works: When someone gets exceptional AI output, capture the prompt that produced it. Create a simple process (a Slack channel, a shared doc, a wiki page) where people can share effective prompts.

Iterate based on feedback: Prompts improve with use. Track which prompts work reliably, which need adjustment, and which should be retired. The best libraries are living documents, not static references.

Include context and instructions: Don't just save the prompt text. Document what it's for, when to use it, what inputs it needs, and any tips for getting the best results. Future users need enough context to use prompts effectively.

The Compounding Effect

A prompt library creates compounding returns. Every prompt saved is time saved for future users. Every refinement improves results for everyone. Every new hire gets the accumulated prompting wisdom of the entire organization instead of starting from scratch.

Companies with mature prompt libraries report that new employees achieve prompting competence in days instead of months. The library becomes institutional knowledge that persists even as people change roles or leave. This is a competitive advantage that grows over time.

Platform-Specific Tips

The universal framework works across all platforms, but each has features worth knowing about.

Platform	Key Features for Business Users
ChatGPT	Custom GPTs save role/context/requirements as reusable assistants. Memory carries context across conversations automatically. Deep research mode runs multi-step investigations with cited reports. Canvas provides interactive editing for writing and code.
Claude	Projects let you set persistent instructions and upload reference documents. Memory carries details between chats. Up to 1M tokens of context for processing large document sets or codebases. MCP integrations connect Claude to external tools and data sources. Built-in web search, code execution, and file creation.
Gemini	Deep Google Workspace integration works natively with Docs, Sheets, Gmail, and Drive. Gems let you create custom personas with saved instructions. Up to 1M tokens of context on Gemini 2.5 Pro. Handles multimodal input including audio and video. NotebookLM offers a separate research tool for uploaded sources.
Copilot	Embedded across Microsoft 365 inside Word, Excel, Outlook, Teams, and PowerPoint. Best for organizations already in the Microsoft ecosystem. Copilot Agents and Copilot Studio let you build custom AI workflows without code. Enterprise security and compliance features for regulated industries.

The honest truth: for most business tasks, any of these platforms will serve you well. The quality of your prompts matters more than your choice of platform.

Don't get wrapped up with tools. Start with whichever tool fits your existing workflow, then focus on developing your prompting skills.

Advanced Techniques

Once you've mastered the basics, these techniques take your prompting to the next level.

Use the Platform's Optimizer

Before manually refining a complex prompt, run it through your platform's built-in optimizer. ChatGPT has prompt optimization in Playground. Claude has a prompt improver in Console. Gemini has optimization in AI Studio. These tools catch common issues automatically and often suggest improvements you wouldn't have thought of.

Request Reasoning Before Conclusions

For analytical tasks, ask AI to show its work before delivering conclusions. "First, list the key factors to consider. Then, analyze each factor. Finally, provide your recommendation with confidence level." This step-by-step approach improves accuracy by forcing systematic analysis rather than jumping to conclusions.

Ask for Self-Verification

For high-stakes outputs, end your prompt with: "After completing your response, verify: (1) all calculations are correct, (2) all claims are supported by the provided data, (3) all requirements above have been addressed, (4) flag any areas of uncertainty." This catches errors that would otherwise slip through.

Define Your Audience

Specify who will read the output and what they already know. "Audience: Board members with finance backgrounds. Assume familiarity with GAAP accounting and our industry metrics. Do not assume familiarity with internal project codenames." This calibrates terminology and depth automatically.

Key Takeaways

1. **Prompting is communication, not coding.** The same skills that make you effective at delegating to humans (clarity, specificity, context) make you effective at working with AI. You're just being more explicit.

2. **Structure matters: Role, Context, Task, Requirements, Output Format, Examples.** Not every prompt needs every element, but knowing the framework helps you diagnose why a prompt isn't working.

3. **Be specific about everything, especially length and format.** Vague prompts produce vague results. Words like "good" and "detailed" are meaningless unless you define them.

4. **Examples are your most powerful tool.** Showing AI what good looks like is more effective than describing it. Two to three examples is usually enough.

5. **Say what NOT to do, not just what to do.** Negative instructions prevent AI's default behaviors that don't match your needs. Think about what typically goes wrong and explicitly prohibit it.

6. **Build organizational capability, not just individual skill.** A prompt library creates compounding returns. Every effective prompt shared benefits everyone. This becomes competitive advantage.

Your 48-Hour Challenge

Before moving to Module 6, complete at least one of these exercises:

1. **Transform a weak prompt:** Take a prompt you've used recently that produced mediocre results. Rewrite it using the six-element framework. Compare the outputs.
2. **Use a template for real work:** Pick one template from this module and use it for an actual task this week. Note what worked and what you'd adjust.
3. **Start your library:** Create a document or folder for saving effective prompts. Add at least three prompts that you or your team use regularly. Include notes on what each is for and tips for best results.
4. **Share a win:** Show a colleague the difference between weak and strong prompting using a real task. Nothing converts skeptics faster than seeing the difference firsthand.

What's Next

You can now write solid prompts and get useful output from AI tools. That's a real skill, and it puts you ahead of most executives.

But I want you to think about something. Every prompt your team sends to ChatGPT contains company information. Every document they upload for summarization goes somewhere. Every customer name they paste into a prompt window is leaving your control. Module 6 is about governance, and if that word sounds boring, think of it as the thing that keeps your AI adoption from turning into a data breach headline.

THE AI OWNER'S MANUAL

AI Governance

Protecting your organization while enabling AI innovation: policies, data management, and security frameworks that actually work.

Module Overview

You've learned what AI can do, where to focus, how to implement it, and how to communicate with AI effectively. But there's a critical challenge most organizations face that we haven't fully addressed: **governance**. The policies, procedures, and technical controls that ensure AI adoption doesn't expose your organization to catastrophic risk.

Here's the uncomfortable reality: most organizations are already using AI, whether leadership knows it or not. Employees are pasting customer data into free or personal ChatGPT accounts. Teams are uploading confidential documents to AI tools with no data protection agreements. Your company's most sensitive information is scattered across SharePoint libraries, Google Drive folders, Dropbox accounts, and file servers and now that messy, unstructured data is now being fed to AI systems that struggle to distinguish current information from outdated drafts or confidential client data from public marketing materials.

This isn't a theoretical risk. Organizations without AI governance are experiencing data breaches, compliance violations, intellectual property leaks, and AI outputs so polluted with outdated information that they're worse than useless. These activities are toxic to decision-making.

This module is the most thorough in our series for a reason. Governance isn't just a checkbox. It's the foundation that determines whether your AI investment creates value or creates liability. We'll cover policy frameworks, data management, technical controls, shadow AI detection, and incident response. You'll leave with actionable frameworks you can implement immediately, regardless of your organization's size or technical sophistication.

Learning Objectives

- ✓ Understand the real risks of ungoverned AI; data exposure, quality degradation, compliance violations, and shadow AI
- ✓ Recognize why your current data management practices are creating AI problems
- ✓ Implement a practical data governance framework designed for AI readiness
- ✓ Deploy an AI acceptable use policy that enables innovation while protecting assets
- ✓ Identify and eliminate shadow AI usage before it causes a breach
- ✓ Establish technical controls that enforce governance automatically
- ✓ Prepare for AI-specific compliance requirements and regulatory changes

The Ungoverned AI Crisis

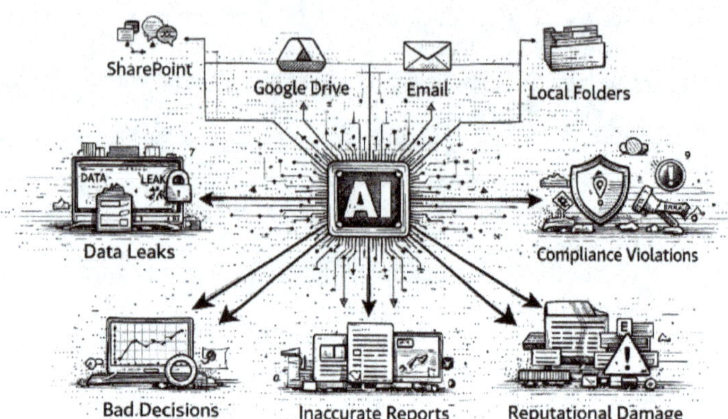

Before we discuss solutions, let's be honest about the scope of the problem. Most companies are operating in a state of ungoverned AI chaos, and the risks compound daily.

Your employees are already using AI, whether you've approved it or not. Your business data is scattered across more systems than anyone wants to admit. And the gap between those two realities is where breaches, compliance failures, and bad decisions quietly take root.

None of it requires panic, but all of it requires attention.

The Data Sprawl Problem

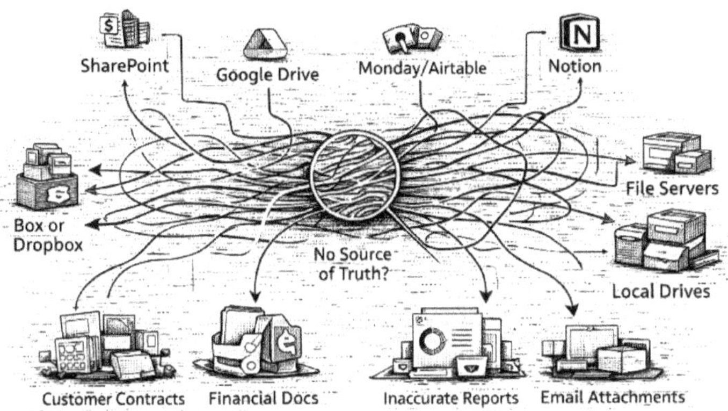

Walk into any organization with 25-500 employees and you'll find the same pattern: business-critical data scattered across a dozen different systems with no unified governance.

Consider a typical mid-sized company's data situation:

- **SharePoint:** Department sites with years of accumulated documents, multiple versions of the same file, abandoned project folders nobody dares delete, and permission structures that have evolved organically into incomprehensible complexity
- **Google Drive:** Some teams use Workspace, creating islands of data disconnected from the Microsoft ecosystem, with sharing links that may or may not still be appropriate for their current audience
- **Box or Dropbox:** Often adopted by individual teams or executives who preferred a different interface, now containing customer contracts and financial documents that exist nowhere else
- **File servers:** The legacy backbone nobody wants to migrate, holding historical records mixed with active

projects, with folder structures that made sense a decade ago but now require institutional knowledge to navigate
- **Email attachments:** The actual source of truth for many important documents, buried in inboxes with no central indexing
- **Local drives:** Desktop folders where employees save "important" files because it's faster than working through the corporate systems
- **Shadow IT applications:** Notion workspaces, Airtable bases, Monday boards, and other tools that teams adopted without IT approval, each containing unique business data

This wasn't as critical of a problem before AI. Employees knew where their stuff was, they asked colleagues when they couldn't find something, and institutional knowledge filled the gaps. Inefficient, yes. Dangerous, not particularly.

AI changes everything.

Now you're asking AI systems to draw on your organizational knowledge to answer questions, generate documents, and inform decisions. And AI doesn't have institutional knowledge. It doesn't know that the "Q3 Forecast Final FINAL v2.xlsx" is actually outdated and you should use "Q3 Forecast - After Board Meeting.xlsx" instead. It doesn't understand that the customer contract in the old project folder has been superseded by an amendment in someone's email. It treats every document it can access as equally authoritative.

The Quality Degradation Spiral

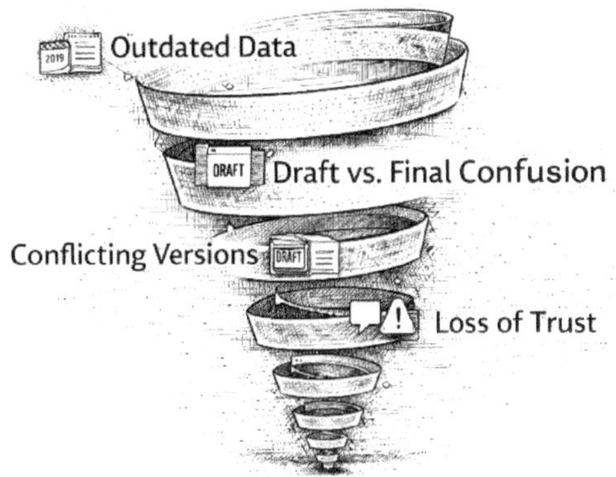

When AI systems ingest unmanaged data, output quality suffers in predictable ways:

1. **Contradictory information:** The AI finds three different versions of your pricing structure and averages them, or maybe it picks one arbitrarily, or hallucinates a fourth option that seems to reconcile the differences.
2. **Outdated data:** Your AI assistant confidently references a policy that was replaced two years ago because the old version still exists in an archive folder with no clear indication that it's obsolete.
3. **Draft vs. final confusion:** AI pulls language from a draft document that was never approved, presenting speculation as company positions.
4. **Context collapse:** Information appropriate for one audience (internal strategy discussions) gets mixed with information for another (customer communications), creating outputs that are tone-deaf or inappropriate.
5. **Permissions inheritance:** If AI has access to HR files, it might reference salary data or performance reviews in

responses that go to people who shouldn't see that information.

Organizations experiencing these issues often blame the AI tool. "It's not accurate." "The outputs aren't useful." In most cases, the problem isn't the AI. It's the data you're feeding it.

The Shadow AI Problem

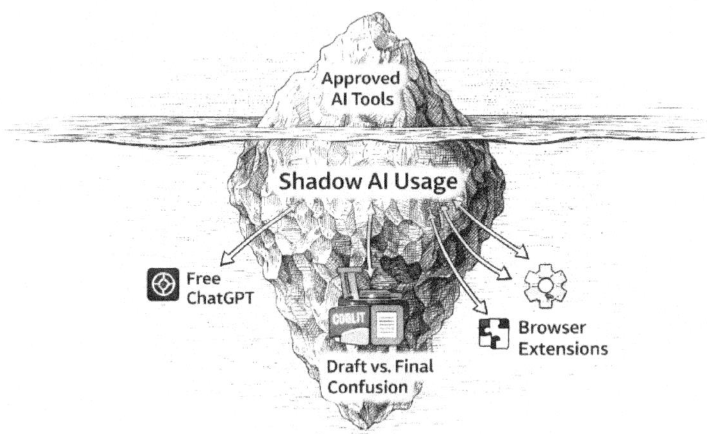

Outdated data isn't you're only issue. Right now, while you're reading this module, your employees are using AI. The question is whether they're using it safely.

Shadow AI refers to unauthorized use of AI tools for business purposes. Typically, employees using free personal accounts to process company data. Recent industry surveys indicate that 70-80% of employees at companies without AI policies have experimented with consumer AI tools for work tasks. Many do so daily.

Common shadow AI scenarios:

- A sales rep pastes a customer's confidential requirements into ChatGPT to generate a proposal faster
- An HR manager uploads resumes with personal information to an AI tool to help screen candidates
- A developer pastes proprietary code into Copilot or another coding assistant without a business license
- A finance employee uploads a spreadsheet with customer payment data to get help with analysis

- An executive copies board meeting notes into Claude to help draft a summary
- A marketing team member uploads competitive intelligence documents to help write analysis

In each case, the employee is trying to be more productive. They're not malicious actors. They're people who want to do better work faster. But free AI tools typically have terms of service that allow the provider to use inputs for model training. That means your customer data, your proprietary processes, your strategic plans may be incorporated into a public AI model that your competitors can access.

The Enterprise vs. Consumer Gap

Consumer vs. Enterprise AI

Consumer AI	Enterprise AI
Training Data Usage — Your inputs used to improve the system	**Training data excluded** unless allowed
Audit Logs — Limited or no audit trail	**Audit Logs** — Limited or no audit trail
Compliance — Little focus on regulatory requirements	**Compliance** — Contractual compliance attestation
Retention — Data stored indefinitely	**Retention** — Data retention configurable

This is where the distinction between enterprise and consumer AI tools is critical:

Feature	Consumer/Free AI	Enterprise AI
Data Training	May use inputs for training	Your data never used for training
Data Retention	May retain conversations	Configurable retention policies
Data Location	Unspecified regions	Specified data residency
Compliance	Consumer privacy only	SOC 2, HIPAA, GDPR options
Access Controls	Individual accounts only	SSO, role-based access, audit logs
Data Protection Agreement	Consumer ToS only	Business associate / DPA available

The cost difference is typically $20-30 per user per month. That's the price of a business lunch, and it's the difference between protected and unprotected data handling. Yet many organizations hesitate on this expense while their employees find workarounds with consumer tools and then unknowingly put the company at risk.

Data Governance Foundations for AI

Effective AI governance starts with data governance. You cannot control AI risks if you don't control your data. This section provides a practical framework for organizations that don't have formal data loss prevention (DLP) systems in place, which is most SMBs.

Data Classification for AI

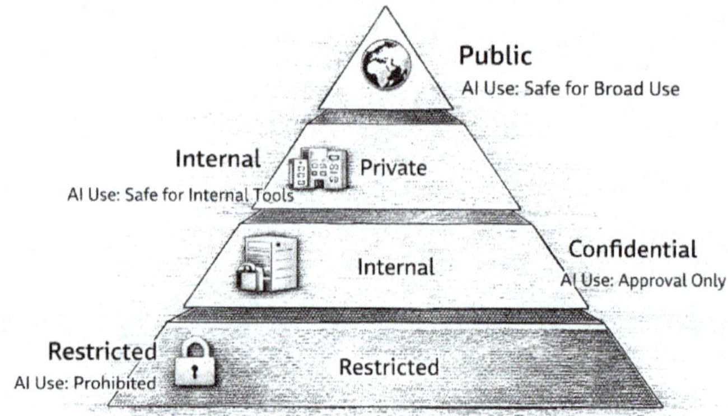

Every piece of data in your organization falls somewhere on a sensitivity spectrum. Before AI can safely use your data, you need to know what you have and how sensitive it is. Here's a practical four-tier classification system designed specifically for AI governance:

Classification	Description & Examples	AI Usage Rules	Access Level
PUBLIC	Published marketing materials, press releases, public website content, published product documentation	Any approved AI tool, no restrictions	All employees
INTERNAL	Meeting notes, internal memos, process documentation, training materials, non-sensitive operational data	Enterprise AI tools with DPA only; never consumer tools	All employees
CONFIDENTIAL	Financial reports, strategic plans, customer lists, pricing structures, unpublished product plans, vendor contracts	Enterprise AI with explicit approval; requires anonymization when possible	Need-to-know basis
RESTRICTED	PII/PHI, credentials, payment data, legal privileged, trade secrets, M&A discussions	No AI usage without specific system approval; most categories prohibited entirely	Named individuals only

Data That Must Never Enter AI Systems

Regardless of the AI tool's security certifications, certain data categories should be strictly prohibited from AI input:

- Social Security numbers, government ID numbers, or financial account numbers

- Passwords, API keys, encryption keys, or any authentication credentials
- Protected health information (PHI) unless using a HIPAA-compliant tool with a BAA
- Payment card data (credit card numbers, CVVs, expiration dates)
- Attorney-client privileged communications
- Unannounced M&A or material nonpublic information
- Biometric data or facial recognition templates

This isn't about paranoia. It's about recognizing that even the most secure AI systems represent an additional attack surface and data processing location.

The security principle of data minimization applies: *don't expose data to systems that don't absolutely need it.*

Tackling the Data Sprawl

Now let's address the practical challenge: you have data everywhere, in various states of organization, and you need to make it AI-ready without disrupting business operations. Here's a phased approach:

Phase 1: Discovery and Mapping (2-4 weeks)
Before you can govern data, you need to know where it is. Conduct a thorough data inventory:

1. **List all data storage systems**: SharePoint sites, Google Drive domains, Box/Dropbox accounts, file servers, NAS devices, individual cloud storage, shadow IT applications. Don't rely on IT's official inventory. Send a survey asking teams where they actually store work documents.
2. **Identify data owners**: Who is responsible for each repository? If no one claims ownership, that's a governance gap that needs immediate attention.
3. **Map sensitive data locations**: Where does customer PII live? Where are financial records? Where are HR files?

Create a heat map of sensitivity across your storage footprint.

4. **Document access patterns**: Who has access to what? Are there over-permissioned shares? Legacy permissions that no longer make sense?

Phase 2: Establish AI-Ready Zones (4-8 weeks)

Rather than trying to clean up everything at once, create designated "AI-ready" areas with proper governance. This allows you to enable AI safely in specific domains while addressing broader data management over time.

1. **Create curated knowledge bases**: Identify high-value content that would benefit from AI access. Copy (not move) authoritative, current versions into dedicated AI-ready repositories. Implement clear naming conventions and version control.
2. **Establish content freshness rules**: Define how content gets into AI-ready zones. Require explicit approval or tagging. Implement expiration dates or review cycles. Mark content with "last verified" dates.
3. **Separate by sensitivity**: Create distinct zones for Public, Internal, and Confidential content. Restricted content stays out entirely. This allows you to connect different AI tools to different zones based on their security capabilities.
4. **Implement metadata standards**: Tag content with classification level, document type, owner, and status (draft/final/superseded). This helps both humans and AI systems understand context.

Phase 3: Address Legacy Data (Ongoing)

The bulk of your existing data won't fit neatly into AI-ready zones, and that's okay. Here's how to manage it:

1. **Default to exclusion:** Don't give AI tools access to unclassified legacy data. Err on the side of caution.

2. **Prioritize high-traffic areas:** Clean up repositories that people actually use. Ignore archives that haven't been touched in years.

3. **Implement retirement policies:** Content older than X years that hasn't been accessed moves to archive status and is excluded from AI indexing.

4. **Use AI to help classify:** Ironically, AI tools can help identify sensitive content in legacy repositories by scanning for patterns that indicate PII, financial data, or other restricted content.

The AI Policy Framework

An AI acceptable use policy is the cornerstone of governance. Done well, it enables employees to use AI productively while protecting the organization. Done poorly (or not at all) it either creates so much friction that people work around it or provides no meaningful protection.

Essential Policy Components

Every AI acceptable use policy needs to address these core areas:

1. Purpose and Scope

Define clearly who the policy applies to (all employees, contractors, third parties) and what it covers (all AI tools, not just the obvious ones). Remember that AI is embedded in many business applications now: email assistants, document summarizers, search enhancements. Your scope should capture these embedded AI features, not just standalone chatbots.

2. Approved Tools and Usage Tiers

List specifically which AI tools are approved for which use cases. A tiered approach works well:

Tier	Tools	Allowed Data	Approval Required
Tier 1: General Use	Microsoft Copilot (Enterprise), Claude Team/Enterprise	Public and Internal data	None - available to all
Tier 2: Supervised Use	Specialized AI tools with DPAs	Confidential with anonymization	Manager + IT approval
Tier 3: Restricted Use	HIPAA/SOC2 certified tools only	Specific regulated data types	Compliance officer approval
Prohibited	Free/consumer AI tools	No company data permitted	N/A- not allowed

3. Data Handling Requirements

Reference your data classification system. Specify what data can go into which tools. Provide concrete examples as employees need to understand what "confidential" means in practice.

- **Give examples of acceptable inputs:** "You may ask Claude to help draft a blog post about industry trends."
- **Give examples of prohibited inputs:** "You may not paste customer names and addresses into any AI tool."
- **Explain the "redaction test":** "Before pasting any content, ask: would this be okay if it appeared in a news article attributed to our company?"

4. Output Review and Accountability

AI outputs are drafts, not final products. Your policy should establish:

- **Human review requirements:** All AI-generated content must be reviewed by a qualified human before external use or consequential decisions.
- **Verification obligations:** Facts and figures generated by AI must be verified against authoritative sources.
- **Attribution standards:** When is disclosure of AI assistance required? This varies by context as academic work, regulatory filings, and creative work may all have different standards.
- **Accountability principle:** The employee who submits AI-assisted work is responsible for it. "The AI told me" is not an acceptable defense.

5. Prohibited Uses

Be explicit about what's not allowed. Common prohibitions include:

- Using AI to make or justify personnel decisions without human judgment

- Generating content that misrepresents human authorship when attribution matters
- Submitting AI-generated work as original in academic or certification contexts
- Bypassing security controls or using AI to generate malicious code
- Creating deepfakes, synthetic media, or misleading content
- Using AI in ways that violate customer contracts or privacy expectations

6. Incident Reporting and Enforcement

Define what constitutes a policy violation and how to report concerns. Establish clear escalation paths. Make consequences proportional. Accidental minor violations shouldn't be treated the same as intentional data breaches. Include a safe harbor for good-faith self-reporting.

Policy Design Principles

A policy that no one follows is worse than no policy. It creates a false sense of security. These principles increase the likelihood of actual adoption:

1. **Enable, don't just restrict.** The policy should make it easy to use AI safely, not just tell people what they can't do. If the approved path is harder than the shadow AI path, people will take the shortcut.
2. **Use concrete examples.** Abstract rules are hard to apply. Specific scenarios make compliance intuitive.
3. **Address the "why."** People comply more readily when they understand the reasoning. Explain the risks, not just the rules.
4. **Plan for evolution.** AI capabilities change rapidly. Build in review cycles and amendment processes. A policy frozen in 2025 will be obsolete by 2027.
5. **Make it findable.** The best policy in the world doesn't help if employees can't find it when they have a question.

Integrate it into existing reference systems and provide a quick reference card.

Detecting and Managing Shadow AI

Shadow AI is already in your organization. The question isn't whether to address it, but how.

Discovery Methods

Finding shadow AI usage requires multiple approaches:

Technical Detection
- **Network traffic analysis:** Your firewall logs show connections to api.openai.com, anthropic.com, and other AI endpoints. Patterns of usage reveal who's accessing what.
- **Browser extension audits:** Many AI tools are accessed via browser extensions. Review installed extensions across managed devices.
- **Expense report analysis:** Employees expensing ChatGPT Plus or other AI subscriptions are self-reporting shadow AI usage.
- **Email patterns:** Automated emails from AI services (account confirmations, usage reports) can indicate personal AI account creation.
- **SaaS management tools:** If you use a SaaS management platform, it may already be tracking AI tool usage alongside other applications.

Human Intelligence
- **Anonymous surveys:** Ask employees directly about their AI usage. Promise no repercussions for honest answers. You might be surprised how forthcoming people are when asked without judgment.
- **Manager conversations:** Team leads often know which team members are AI enthusiasts. Frame it as discovery, not enforcement.
- **Output pattern recognition:** Sudden changes in writing style, dramatic productivity increases, or outputs that feel "too polished" may indicate AI assistance.

The Conversion Strategy

The goal isn't to punish shadow AI users. It's to convert them to sanctioned tools. These employees are your AI pioneers. They're demonstrating demand and developing skills that benefit the organization. You want to channel that energy, not suppress it.

Step 1: Acknowledge the Value

When you identify shadow AI usage, start by acknowledging that the employee was trying to be more productive. Thank them for their initiative. Explain that you're not there to punish. You're there to make sure they can continue using AI safely.

Step 2: Understand the Use Case

Ask what they're using AI for. This intelligence is valuable. It shows you where AI is providing real value in your organization. It reveals gaps in your official tool offerings. It identifies use cases you might not have considered.

Step 3: Provide a Better Alternative

Show them how to accomplish the same tasks with approved tools. If you don't have an approved alternative that meets their needs, that's feedback for your governance program and not a failure by the employee. Consider fast-tracking approval for tools that address demonstrated demand.

Step 4: Remove the Need for Workarounds

Make official AI tools easy to access. Reduce approval friction. Ensure licenses are available without bureaucratic delays. If employees can start using sanctioned tools in minutes, there's no incentive to maintain shadow accounts.

When Enforcement Is Necessary

Sometimes shadow AI usage does require enforcement action. Particularly when sensitive data has been exposed. Calibrate your response to the severity:

Scenario	Risk Level	Appropriate Response
Employee used free AI for personal productivity (no sensitive data)	Low	Education, transition to approved tools
Employee uploaded internal (non-confidential) documents	Medium	Education, policy reminder, documentation
Employee uploaded customer data or PII	High	Formal incident, possible breach notification, disciplinary action
Repeated violations after warnings	High	Progressive discipline up to termination
Intentional circumvention of controls	Critical	Immediate access revocation, serious disciplinary action

Technical Controls for AI Governance

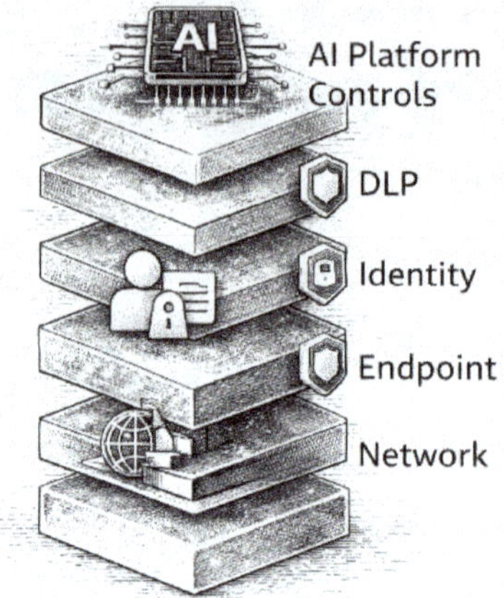

Policy without enforcement is just wishful thinking. Technical controls make governance automatic, reducing reliance on employee awareness and good intentions.

Network-Level Controls

Web Filtering and Categorization

Configure your web filter or secure web gateway to categorize AI services and apply appropriate policies:

- **Block consumer AI sites:** chat.openai.com (consumer version), free tiers of Claude, Gemini, etc. This is aggressive but effective for high-security environments.
- **Allow-list enterprise endpoints:** Explicitly permit your sanctioned AI tools while blocking unapproved alternatives.
- **Monitor without blocking:** For organizations preferring visibility over enforcement, log all AI site access for review.

- **Warning pages:** Interstitial pages that remind users of policy before allowing access can provide awareness without hard blocking.

DNS-Level Filtering

DNS filtering services (Cisco Umbrella, Cloudflare Gateway, Zscaler) can block AI domains at the network level, providing coverage even for unmanaged devices on your network. This is particularly valuable for guest WiFi and BYOD scenarios.

Endpoint Controls

Browser Management

For managed devices, browser policies provide granular control:

- Block AI-related browser extensions that haven't been vetted
- Enforce enterprise profiles that separate work and personal browsing
- Deploy browser-based DLP extensions that scan content before it leaves the browser
- Configure enterprise AI tools as managed browser extensions with automatic deployment

Application Control

Beyond web browsers, AI tools may come as desktop applications, mobile apps, or IDE plugins:

- **Software inventory:** Regularly scan for AI-related applications installed on managed devices.
- **Application allow-listing:** In high-security environments, only permit pre-approved applications to run.
- **Mobile device management:** Apply similar policies to company mobile devices, including app store restrictions.

Data Loss Prevention (DLP) Integration

If you have a DLP solution, extend it to cover AI tools:

- Create policies that monitor or block sensitive data patterns (SSNs, credit card numbers, health records) being sent to AI endpoints
- Configure alerts for bulk document uploads to AI services
- Monitor clipboard activity for sensitive data being pasted into web applications
- Integrate DLP alerts with your security operations workflow

If you don't have enterprise DLP, consider lighter-weight options:

- Microsoft Purview (included in many M365 licenses) provides basic DLP capabilities
- Google Workspace has built-in DLP rules for Workspace data
- Browser-based DLP extensions can provide coverage without full endpoint agents

Identity and Access Management

Single Sign-On Integration

Require SSO authentication for all sanctioned AI tools. This provides:

- **Centralized access control:** Provision and deprovision access from your identity provider
- **Audit trail:** Know who accessed which AI tools and when
- **MFA enforcement:** Apply your authentication policies consistently
- **Automatic offboarding:** When employees leave, their AI access terminates with their other system access

Conditional Access Policies

Apply conditional access rules to AI tools:

- Require managed devices for access to AI tools that might see sensitive data
- Restrict AI access to specific IP ranges or require VPN connection
- Apply different access levels based on user groups or sensitivity of the AI application
- Block access from regions where your organization doesn't operate

AI-Specific Security Configurations

Enterprise AI tools offer configuration options that affect security. Make sure you've properly configured:

- **Data retention settings:** How long does the AI provider keep your conversation data? Set the minimum necessary.
- **Training opt-out:** Ensure your data isn't used to train public models.
- **Admin controls:** Configure which features users can access, what integrations are permitted, and what data sources can be connected.
- **Audit logging:** Enable thorough logging of user activities for compliance and incident investigation.
- **Data residency:** If available, specify where your data is processed and stored.

This is where many businesses struggle: properly planning, implementing, and managing data protection, classification, and governance tools. That's where working with a trusted advisor and technology partner like QIT Solutions becomes essential.

Compliance and Regulatory Considerations

AI governance doesn't exist in a regulatory vacuum. Existing compliance obligations apply to AI usage, and new AI-specific regulations are emerging rapidly.

Existing Regulations Applied to AI

Data Protection Regulations

GDPR, CCPA, and similar privacy laws apply when AI processes personal data:

- **Legal basis:** Do you have a lawful basis for processing personal data through AI? Consent, legitimate interest, and contract performance are common bases, but they require analysis.
- **Data minimization:** Are you only processing personal data that's necessary for the AI use case?
- **Cross-border transfers:** Where is the AI vendor processing your data? Are appropriate transfer mechanisms in place?
- **Vendor agreements:** Do your AI vendor contracts include required data protection terms?
- **Transparency:** Does your privacy notice inform individuals that their data may be processed by AI systems?

Industry-Specific Compliance

Regulation	AI Implications	Key Requirements
HIPAA	PHI in AI requires BAA with vendor; AI counts as a "business associate"	Signed BAA, minimum necessary standard, breach notification procedures
SOC 2	AI vendors handling customer data should be SOC 2 compliant	Request SOC 2 Type II reports from AI vendors; ensure controls cover AI-specific risks
PCI DSS	Payment card data should never enter AI systems unless specifically certified	Complete prohibition recommended; if unavoidable, full PCI DSS scoping
SEC/FINRA	AI-generated communications may be subject to recordkeeping and supervision requirements	Retain AI interactions as business records; review and supervision procedures
State Laws	Colorado, Illinois, and others have specific AI requirements	Monitor evolving state laws; some require disclosure of AI use in decisions

Emerging AI-Specific Regulations

The regulatory landscape for AI is evolving rapidly:

- **EU AI Act:** The most thorough AI regulation globally. Creates risk categories for AI systems with corresponding obligations. Organizations with EU presence should be assessing implications now.
- **US Executive Order on AI:** While focused on federal government use, signals direction for eventual private sector requirements. Safety testing, transparency, and civil rights protections are key themes.
- **State AI laws:** Colorado, Utah, and others have passed or are considering AI-specific legislation. Expect a patchwork of requirements similar to the privacy law landscape.
- **Industry self-regulation:** NIST AI Risk Management Framework provides voluntary guidance. Following recognized frameworks demonstrates due diligence even without legal mandate.

Vendor Due Diligence

Before approving any AI tool for business use, conduct proper vendor evaluations:

Minimum Requirements for Any AI Vendor

- Data protection agreement (DPA) or similar contractual commitments
- Clear data retention and deletion policies
- Confirmation that customer data is not used for model training
- Security certifications appropriate to data sensitivity (SOC 2 minimum for business data)
- Incident response and breach notification commitments
- Acceptable sub-processor list (who else touches your data)

Enhanced Requirements for Sensitive Data

- Industry-specific compliance certifications (HIPAA BAA, PCI compliance)
- Data residency guarantees with specified processing locations
- Customer-managed encryption keys option
- Enhanced audit rights and reporting
- Dedicated instance or tenant isolation options

AI Incident Response

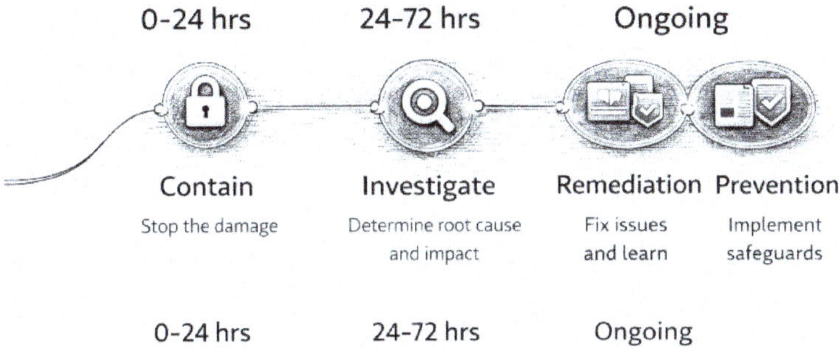

Even with strong governance, AI incidents will occur. Preparation enables effective response.

Types of AI Incidents

AI-related incidents fall into several categories:

1. **Data exposure:** Sensitive data input to an AI system that shouldn't have received it. This includes both intentional misuse and accidental exposure.
2. **Output failures:** AI generates inappropriate, incorrect, or harmful content that reaches customers or influences decisions.
3. **Privacy violations:** AI surfaces personal information in contexts where it shouldn't or it makes decisions using data inappropriately.
4. **Vendor breaches:** An AI vendor experiences a security incident affecting customer data.
5. **Policy violations:** Employees using AI in ways that violate organizational policy, discovered after the fact.

Response Framework

Immediate Response (0-24 hours)

a) **Contain:** Stop the immediate harm. If data was exposed, attempt to delete it from the AI system (many enterprise tools support conversation deletion). Revoke access if necessary.

b) **Document:** Capture what happened, when, and who was involved. Screenshot evidence. Record the specific AI tool and conversation.

c) **Assess severity:** What data was involved? What are the potential consequences? Is this a reportable breach under any applicable regulations?

d) **Notify stakeholders:** Inform legal, compliance, and appropriate leadership based on severity. Don't wait until you have complete information for high-severity incidents.

Investigation (24-72 hours)

e) **Scope the incident:** How much data was affected? Was it a single incident or pattern? Are others doing the same thing?

f) **Interview involved parties:** Understand the circumstances. Was this intentional or accidental? What was the person trying to accomplish?

g) **Vendor engagement:** If data was exposed to an AI provider, contact them. Understand their data handling, request deletion confirmations, document their response.

h) **Legal assessment:** Determine if breach notification requirements apply. Consider engaging breach counsel for significant incidents.

Remediation and Prevention

i) **Root cause analysis:** Why did this happen? Was it a policy gap, training failure, technical control gap, or intentional circumvention?

j) **Control improvements:** What changes would prevent recurrence? Implement them with urgency proportional to risk.

k) **Communication:** If others might make the same mistake, communicate learnings without identifying individuals. Use incidents as teaching moments.

l) **Document lessons learned:** Update policies, training, and controls based on what you learned.

Implementation Roadmap

Implementing thorough AI governance doesn't happen overnight. Here's a practical phased approach:

Phase 1: Foundation (Weeks 1-4)

Focus: Establish basic governance and address immediate risks

- Draft and implement basic AI acceptable use policy
- Identify and approve 1-2 enterprise AI tools for general use
- Communicate policy and provide basic training
- Conduct initial shadow AI discovery
- Establish incident reporting channel
- Document data classification framework

Phase 2: Expansion (Weeks 5-12)

Focus: Extend controls and address data governance

- Complete data inventory and mapping
- Create AI-ready data zones in primary repositories
- Implement network-level controls (blocking, monitoring)
- Integrate AI tools with SSO and identity management
- Evaluate and approve additional AI tools based on demand
- Deploy role-specific training for high-use groups
- Develop vendor due diligence checklist and process

Phase 3: Maturation (Months 4-6)

Focus: Operationalize governance and continuous improvement

- Implement DLP integration for AI endpoints
- Establish regular policy review cycle
- Create AI governance committee or assign ongoing ownership
- Develop metrics and reporting for AI usage and incidents

- Conduct tabletop exercises for AI incident scenarios
- Plan for regulatory changes (EU AI Act, state laws)
- Address remaining legacy data governance gaps

Governance Maturity Model

Use this model to assess where you are and where you need to be:

Level	Policy	Technical Controls	Data Governance
Level 0: None	No AI policy exists	No controls on AI usage	No classification or data zones
Level 1: Basic	Basic acceptable use policy published	Some network monitoring	Basic data classification defined
Level 2: Developing	Detailed policy with approved tools list	SSO integration, some blocking	AI-ready zones established
Level 3: Established	thorough policy with training program	Full endpoint and network controls	Active data governance with metadata
Level 4: Optimized	Continuous policy evolution, culture of compliance	DLP integration, automated enforcement	Enterprise-wide governance, AI-assisted classification

Most organizations should aim to reach Level 2 within 90 days, Level 3 within 6 months. Level 4 represents mature organizations with dedicated governance resources.

Key Takeaways

1. **Your data sprawl is an AI quality problem.** AI tools can't distinguish current from outdated, approved from draft, or appropriate from restricted. Clean data governance is a prerequisite for AI value, not a nice-to-have.

2. **Shadow AI is already happening.** Don't assume your employees are waiting for official approval. Discover what's already being used, convert users to sanctioned tools, and make the official path easier than the workaround.

3. **Enterprise AI tools are worth the cost.** The $20-30 per user per month for enterprise AI subscriptions isn't an expense, it's insurance. Consumer tools may use your data for training, lack audit capabilities, and create compliance exposure.

4. **Policy enables, not just restricts.** The best AI policies make it easy to use AI safely. If your policy is purely restrictive, people will work around it. Balance protection with enablement.

5. **Technical controls enforce policy automatically.** Don't rely solely on employee awareness. Network filtering, SSO integration, and DLP provide enforcement that doesn't depend on good intentions.

6. **Compliance requirements apply to AI.** HIPAA, GDPR, PCI, and industry-specific regulations don't have AI exceptions. Ensure your AI governance addresses existing compliance obligations and prepares for emerging AI-specific requirements.

7. **Prepare for incidents now.** AI incidents will happen. Having a response plan, clear escalation paths, and

documented procedures reduces impact and demonstrates due diligence.

8. **Governance evolves with technology.** AI capabilities change rapidly. Build review cycles into your governance program. Today's controls may be inadequate for tomorrow's AI features.

Your 48-Hour Challenge

Before moving to Module 7, complete at least one of these exercises:

1. **Shadow AI discovery:** Ask your IT team to check firewall logs for connections to OpenAI, Anthropic, and Google AI endpoints over the past 30 days. The volume may surprise you.
2. **Data sprawl audit:** Pick one department and list every location where they store documents. Include official systems and unofficial ones. Map where sensitive data lives.
3. **Policy review:** If you have an AI policy, test it with a scenario: "Can I use ChatGPT to summarize these customer feedback emails?" If the answer isn't clear, your policy needs work.
4. **Vendor evaluation:** Pick one AI tool you're considering and check: Do they have a DPA? What's their training data policy? Do they have SOC 2? Document your findings.

What's Next

If you've done the 48-Hour Challenge and actually looked at what your team is already putting into AI tools, you probably had a moment of "oh, that's more than I thought." That reaction is exactly why governance matters. It's not bureaucracy. It's protection for the investment you're about to make.

Speaking of investment, that's the next conversation. Module 7 is about money: what AI actually costs, how to calculate whether it's paying off, and when the honest answer is "not yet."

THE AI OWNER'S MANUAL

AI Economics for SMBs

Budget confidently, measure real ROI, and build business cases that get approval using frameworks that reflect how AI actually costs and delivers value

Module Overview

You've learned what AI can do, where to start, how to implement it, how your leadership team should use it, how to communicate with AI effectively, and how to govern it responsibly. But there's one question that determines whether any of this actually happens:

Can we afford it? And will it pay off?

AI economics is where many SMB initiatives stall. Not because the technology doesn't work (we've already established it does) but because the financial picture is unclear. Vendor pricing is opaque. ROI claims seem inflated. Hidden costs emerge after commitment. And building a business case that satisfies a skeptical CFO or board requires more than "everyone else is doing it."

This module gives you the financial frameworks you need. We'll cover what AI actually costs (not the marketing version), how to calculate ROI that holds up to scrutiny, the hidden costs that surprise most organizations, when to walk away from an AI investment, and how to build business cases that get approved.

The goal isn't to make you an AI finance expert. It's to give you enough fluency to make confident decisions, ask the right questions of vendors, and justify your AI investments to stakeholders who want evidence, not enthusiasm.

Learning Objectives

- ✓ Understand the real cost structure of AI tools, the subscriptions, usage, integration, and ongoing expenses
- ✓ Apply the Build vs. Buy vs. Subscribe framework to make smarter procurement decisions
- ✓ Calculate AI ROI using methods that reflect actual business value, not vendor projections
- ✓ Identify and budget for hidden costs that catch most organizations off guard
- ✓ Recognize when NOT to invest in AI. The scenarios where the economics don't work
- ✓ Build business cases that get stakeholder buy-in with evidence-based arguments

What AI Actually Costs

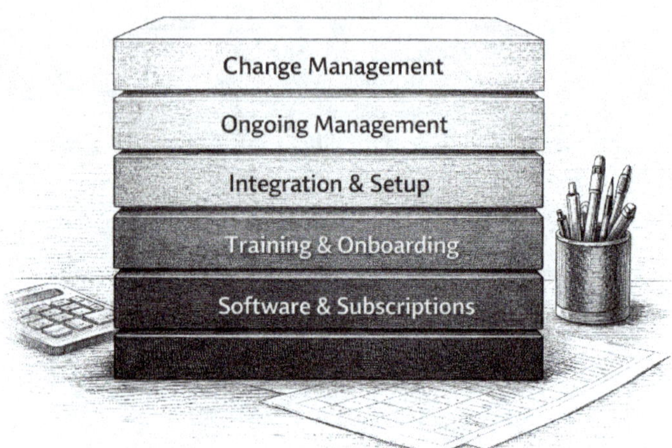

Before we discuss ROI, we need clarity on the cost side of the equation. AI pricing has matured significantly, but it remains more complex than traditional software. Understanding the cost structure helps you budget accurately and negotiate effectively.

What AI Actually Costs

AI tools generally follow one of four pricing models, each with different implications for budgeting and scaling.

Per-Seat Subscriptions

The most familiar model: you pay a fixed monthly fee per user. This is how most enterprise AI tools are sold. Claude for Business, ChatGPT Enterprise, Microsoft Copilot, and similar products. Costs typically range from $20-30 per user per month for general-purpose AI assistants, though specialized tools can run significantly higher.

Advantages: Predictable costs, easy budgeting, usually includes governance features. You know exactly what you'll spend each month.

Disadvantages: Can become expensive at scale if not everyone uses it actively. A 100-person company paying $25/user/month spends $30,000 annually worth it if everyone's productive, wasteful if adoption is patchy.

Usage-Based Pricing

You pay based on how much you use. Typically measured in tokens (roughly words), API calls, or processing time. This is common for developer-facing AI services and custom integrations.

Advantages: You only pay for what you use. Great for variable workloads or early experimentation.

Disadvantages: Unpredictable costs can spike. A process that works fine in testing might cost 10x what you expected at production volume. Requires careful monitoring.

Hybrid Models

Many vendors combine approaches: a base subscription with usage limits, then overage charges. Or tiered pricing where you prepay for a usage bucket. This is increasingly common as vendors try to balance predictability with value capture.

Advantages: Some cost predictability with flexibility for heavy users.

Disadvantages: Complex to model. Easy to underestimate usage and face surprise bills. Requires understanding the tier structure deeply.

One-Time or Project-Based

Some AI implementations (particularly custom development or specialized integrations) are priced as projects. You pay for development, with separate ongoing maintenance fees.

Advantages: Clear project scope, defined deliverables, capital expense treatment.

Disadvantages: Higher upfront costs. Ongoing maintenance and updates often underestimated. Risk of building something that needs replacement as AI evolves.

Realistic Cost Ranges for SMBs

Let's put real numbers on typical AI investments for organizations with 25-500 employees. These ranges reflect 2025 market pricing and should be used as benchmarks when evaluating options.

Category	Annual Cost Range	What's Included
General AI Assistants	$240-$360/user	ChatGPT Enterprise, Claude for Business, Gemini for Workspace. Full-featured enterprise AI.
Productivity Suite AI	$360-$480/user	Microsoft Copilot, Google Workspace AI. Integrated into existing productivity tools.
Specialized Tools	$1,000-$10,000+	Industry-specific AI, advanced analytics, custom models. Pricing varies widely.
Custom Development	$25,000-$250,000+	Custom integrations, proprietary models, bespoke solutions. One-time plus ongoing maintenance.

Example Budget: A 75-person company deploying general AI assistants to 50 knowledge workers, plus Microsoft Copilot for 20 power users, plus one specialized tool for their sales team might budget: $50 \times \$300 + 20 \times \$420 + \$5,000 = \$28,400$ annually in software costs alone. That's before training, integration, or ongoing management.

The Build vs. Buy vs. Subscribe Decision

	Subscribe	Buy	Build
Time to Value	Low	Medium	High
Upfront Cost	Low	Medium	High
Customization	Low	Medium	High
Internal Expertise	Low	Medium	High
Risk Level	Low	Medium	High

One of the most consequential AI decisions is how you acquire capabilities. Each approach has different cost profiles, risk levels, and strategic implications.

Subscribe: Using Off-the-Shelf AI Tools

Subscribing to existing AI services, like enterprise ChatGPT, Claude, or specialized SaaS tools, is the default choice for most SMBs, and usually the right one.

When to Subscribe:

- Your use case is common (writing, research, analysis, coding assistance)
- You need to move quickly (weeks, not months)
- You lack internal AI/ML expertise
- The tool handles governance and security for you
- You want the vendor to handle model updates and improvements

Total Cost of Ownership: Subscription cost + training time + process change management. Usually the lowest TCO option for standard use cases.

Buy: Licensed or Integrated Solutions

Buying involves licensing an AI platform or purchasing an integration that connects AI to your existing systems, like adding AI capabilities to your CRM or ERP.

When to Buy:

- You need AI tightly integrated with existing business systems
- Your use case requires industry-specific capabilities
- You want a proven solution without building from scratch
- Compliance or regulatory requirements demand specific certifications

Total Cost of Ownership: License fees + implementation services + integration costs + ongoing maintenance + training. Typically 2-5x the first-year license cost over three years.

Build: Custom AI Development

Building means creating custom AI capabilities. Your own models, proprietary integrations, or unique applications that don't exist in the market.

When to Build:

- Your use case is truly unique and provides competitive advantage
- You have proprietary data that creates differentiation
- Off-the-shelf options don't meet your requirements
- You have (or can hire) the technical expertise to build and maintain
- The strategic value justifies the investment

Total Cost of Ownership: Development costs + infrastructure + ongoing maintenance + talent retention + iteration and improvement. Often 5-10x initial development cost over three years. Most SMBs underestimate this significantly.

Decision Framework

Use this framework to guide your build/buy/subscribe decision:

Factor	Subscribe	Buy	Build
Time to Value	Days to weeks	Weeks to months	Months to years
Upfront Cost	Low	Medium	High
Customization	Limited	Moderate	Full
Internal Expertise Needed	Low	Medium	High
Vendor Dependency	High	Medium	Low
Best For	Most SMBs, standard use cases	Integration-heavy scenarios	Unique competitive advantage

The Small Business Default: For 90% of AI applications at small and medium businesses, "subscribe" is the right answer. The AI market is mature enough that off-the-shelf tools handle most business needs well. Reserve "buy" for tight system integration and "build" for genuine competitive differentiation.

ROI Calculation Methods That Actually Work

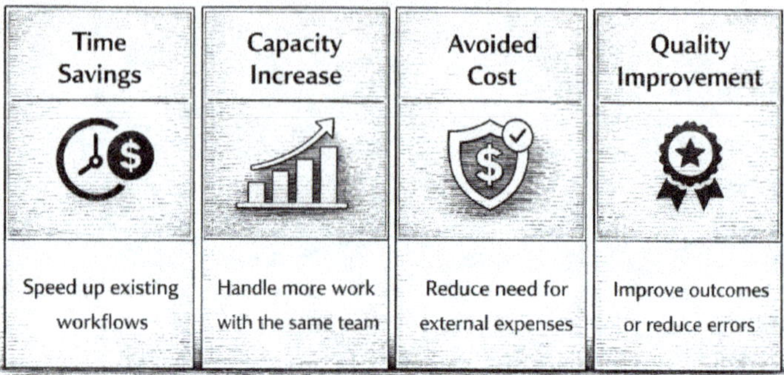

Vendor ROI claims are notoriously inflated. "10x productivity gains" and "500% ROI" make great marketing but rarely reflect reality. To make sound investment decisions, you need calculation methods that produce defensible numbers.

The Time Savings Method

The most straightforward ROI approach: measure how much time AI saves, convert that to labor cost, and compare to AI cost.

Formula:
ROI = (Hours Saved × Hourly Labor Cost × Number of Users) ÷ Total AI Cost

Example: A 50-person company deploys AI writing assistants. Users report saving 5 hours per week on average. Fully-loaded labor cost is $50/hour. AI costs $25/user/month.

- Weekly savings per user: 5 hours × $50 = $250
- Monthly savings per user: $250 × 4 = $1,000
- Company monthly savings: $1,000 × 50 = $50,000
- Company monthly AI cost: $25 × 50 = $1,250
- Monthly ROI: ($50,000 - $1,250) ÷ $1,250 = 3,900%

Reality Check: This math looks amazing, and vendors love to present it this way. But be skeptical. "Saved time" often doesn't convert directly to value. Employees don't necessarily do more billable work with the time saved. A more conservative approach: assume only 25-50% of saved time creates real value.

Conservative ROI: At 33% value capture: ($50,000 × 0.33 - $1,250) ÷ $1,250 = 1,220%. Still excellent, and far more believable.

The Capacity Method

Instead of valuing saved time, measure increased output without headcount growth.

Formula:
ROI = (Additional Output Value - AI Cost) ÷ AI Cost

Example: A marketing team uses AI to increase content production from 20 to 35 pieces per month without adding staff. Each piece generates an average of $500 in attributed revenue.

- Additional monthly output: 15 pieces
- Additional monthly value: 15 × $500 = $7,500
- Monthly AI cost: $500 (team tools)
- Monthly ROI: ($7,500 - $500) ÷ $500 = 1,400%

Why This Works: This method ties AI to concrete business outcomes. It's harder to game because you're measuring actual output, not reported time savings.

The Avoided Cost Method

Calculate what you would have spent to achieve the same outcome without AI: typically additional headcount, overtime, or outsourcing.

Formula:
ROI = (Avoided Cost - AI Cost) ÷ AI Cost

Example: A customer service team handles 30% more tickets without hiring additional staff. They would have needed two new hires at $60,000 each to handle the growth.

- Avoided annual cost: 2 × $60,000 = $120,000
- Annual AI cost: $15,000
- Annual ROI: ($120,000 - $15,000) ÷ $15,000 = 700%

Why This Works: Avoided headcount is a real budget line item that finance teams understand. It's concrete, verifiable, and speaks directly to cost management.

The Quality Improvement Method

Some AI value comes not from speed but from quality: fewer errors, better decisions, reduced rework.

Formula:

ROI = (Error Cost Reduction + Rework Savings) ÷ AI Cost

Example: An accounting team uses AI to review invoices before sending. Error rate drops from 3% to 0.5%, eliminating an average of 25 error-correction cycles per month that each cost $200 in staff time and customer relationship damage.

- Monthly error reduction: 25 - 4 = 21 incidents
- Monthly savings: 21 × $200 = $4,200
- Monthly AI cost: $300
- Monthly ROI: ($4,200 - $300) ÷ $300 = 1,300%

Hidden Costs That Catch Organizations Off Guard

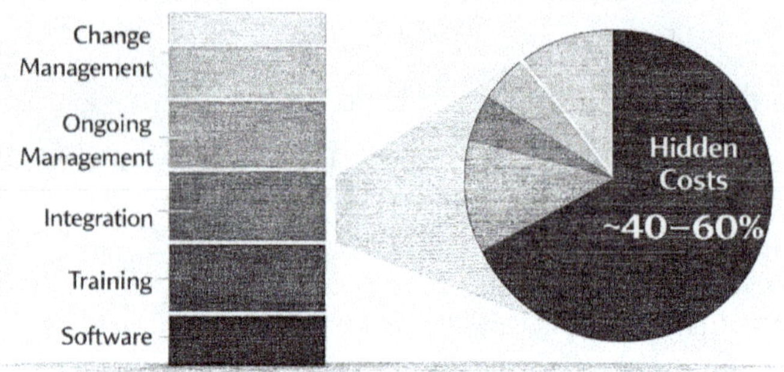

Hidden Costs = ~40–60%

The subscription price is just the beginning. Organizations that budget only for software licenses typically underestimate true AI costs by 40-60%. Here are the expenses that surprise most teams.

Training and Onboarding

AI tools are only valuable if people use them well. Training isn't just a nice-to-have, it's essential for ROI.

- **Initial training time:** 2-8 hours per employee depending on tool complexity
- **Ongoing skill development:** 1-2 hours per month for prompt improvement and new feature adoption
- **External training resources:** $0 (self-serve) to $5,000+ (custom workshops)
- **Productivity dip during learning curve:** Expect 2-4 weeks of reduced output as users adapt

Budget Guidance: Plan for 5-10% of software cost for training in year one, 2-5% in subsequent years.

Integration and Setup

Getting AI tools connected to your existing systems takes time and often money.

- **SSO/identity integration:** 4-20 hours of IT time, potentially vendor professional services
- **Data connections:** Connecting AI to your documents, CRM, or other data sources can range from simple (hours) to complex (weeks)
- **Custom workflow development:** If you want AI embedded in existing processes, expect development time
- **Professional services:** Many vendors offer implementation packages at $5,000-$50,000+

Budget Guidance: For subscription tools, budget 1-2x the first year subscription cost for setup. For "buy" solutions, budget 50-100% of license cost for implementation.

Ongoing Management

AI tools require ongoing attention, they're not set-and-forget.

- **Administration:** User management, policy updates, usage monitoring: 2-5 hours per week
- **Prompt library maintenance:** Developing, testing, and sharing effective prompts: 1-3 hours per week
- **Vendor management:** Staying current on updates, evaluating new features, license optimization
- **Support and troubleshooting:** Helping users with issues, escalating to vendors

Budget Guidance: Plan for 0.1-0.25 FTE dedicated to AI tool management per 50 users. This often falls on IT or a designated "AI champion."

Change Management

Getting people to actually change how they work is harder than buying software.

- **Communication and rollout:** Internal marketing, launch events, ongoing reinforcement
- **Process documentation:** Updating SOPs, creating guidelines, documenting best practices
- **Resistance management:** Time spent addressing concerns, working with skeptics, building buy-in
- **Executive sponsorship:** Leadership time for modeling, advocacy, and accountability

Budget Guidance: This is often unbudgeted and underestimated. Plan for 10-20% of total project cost for change management activities.

Total Cost of Ownership Example

Here's what a realistic first-year budget might look like for a 50-person company deploying enterprise AI assistants:

Cost Category	Amount	% of Total
Software subscriptions (50 users × $25/mo × 12)	$15,000	52%
Initial training (4 hrs × 50 users × $50/hr)	$10,000	35%
IT setup and integration (20 hrs × $75/hr)	$1,500	5%
Ongoing management (0.1 FTE × $80K)	$8,000	28%
TOTAL FIRST-YEAR COST	**$28,900**	**100%**

Key Insight: The "hidden" costs (training, setup, management) nearly double the software-only cost. Year two will be lower (roughly $18,000-20,000 as one-time costs drop off), but year one TCO is nearly twice the subscription price.

When NOT to Invest in AI

Not every AI investment makes sense. Knowing when to say "not now" or "not this" is as important as knowing when to proceed. Here are scenarios where the economics typically don't work.

Your Process Is Broken

AI amplifies existing processes, it doesn't fix them. If your current workflow is inefficient, disorganized, or poorly documented, adding AI will create AI-powered chaos.

Red Flags:

- No clear process documentation exists
- Different team members do the same task completely differently
- Quality is inconsistent even without AI
- There's no clear definition of "good" output

Better Investment: Process improvement and standardization first. Then AI.

The Volume Doesn't Justify It

AI tools have setup costs and learning curves. If you don't do enough of a task to justify the overhead, the ROI won't materialize.

Red Flags:

- The task happens less than weekly
- Only one or two people would use the tool
- Time savings don't cover the subscription cost

- The use case is too narrow to justify training investment

Better Investment: Focus AI on high-volume, repetitive tasks. Handle low-volume tasks manually or through existing tools.

You Can't Measure Success

If you can't define and measure what success looks like, you can't calculate ROI and you can't course-correct if things aren't working.

Red Flags:

- "We want to explore AI" without a specific goal
- No current baseline metrics for the process
- Success is defined vaguely as "being more innovative"
- No plan for tracking AI impact

Better Investment: Establish baselines and success metrics first. Then you'll know if AI is delivering value.

Your Data Isn't Ready

Module 6 covered data governance for a reason. If your data is scattered, inconsistent, or inaccessible, AI tools will struggle to deliver value.

Red Flags:

- Documents are scattered across multiple unconnected systems
- No clear distinction between current and outdated information
- Sensitive data isn't classified or protected

- You can't easily share data with AI tools due to format or access issues

Better Investment: Data consolidation and governance first. AI becomes significantly more valuable with clean, organized data.

The Organization Isn't Ready

Sometimes the technology is ready but the people aren't. Cultural resistance, competing priorities, or lack of leadership support can doom even well-designed AI initiatives.

Red Flags:

- Leadership is pushing AI without personal commitment to using it
- The team is already overwhelmed with other change initiatives
- There's active resistance or significant fear about AI
- No one has time allocated to champion the implementation

Better Investment: Address organizational readiness first. A small, successful pilot with willing participants beats a forced rollout that breeds resentment.

Building the Business Case

Whether you need board approval, CFO sign-off, or partner buy-in, you'll need a compelling business case. Here's how to build one that gets approved.

Know Your Audience

Different stakeholders care about different things. Tailor your case accordingly.

- **CFO/Finance:** ROI, payback period, budget impact, risk mitigation
- **CEO/Board:** Competitive positioning, strategic alignment, growth enablement
- **COO/Operations:** Efficiency gains, capacity increases, quality improvements
- **IT/Security:** Integration complexity, security controls, compliance requirements
- **HR/Team Leads:** Employee impact, training needs, change management

The Business Case Structure

A strong business case follows a logical structure that addresses both opportunity and risk.

1. Executive Summary

One paragraph that captures: what you want to do, why it matters, what it costs, and what you expect in return. Busy executives may read only this.

2. Problem/Opportunity Statement

What business problem are you solving? Frame it in business terms, not technology terms. "Our proposal development takes 40 hours per project, limiting how many opportunities we can pursue" is better than "We need AI to write proposals."

3. Proposed Solution

What specifically will you implement? Be concrete: which tool, for which users, for which use cases. Include the build/buy/subscribe rationale.

4. Financial Analysis

This is where your ROI calculations go. Include total cost of ownership (not just software), expected benefits with conservative and optimistic scenarios, payback period, and comparison to alternatives (including doing nothing).

5. Risk Assessment

What could go wrong? Address implementation risks, adoption risks, security risks, and vendor risks. For each risk, include probability, impact, and mitigation strategy. Acknowledging risks builds credibility.

6. Implementation Plan

How will you roll this out? Include timeline, resource requirements, key milestones, and success metrics. Reference Module 3's 90-day framework if applicable.

7. Recommendation and Ask

Be explicit about what you're asking for: budget amount, approval to proceed, resources needed. Make the decision easy by being specific.

Business Case Best Practices

1. **Use conservative numbers.** If your case works with pessimistic assumptions, it will definitely work with realistic ones. Overpromising destroys credibility.
2. **Show your math.** Don't just state ROI, show how you calculated it. Transparency builds trust and allows stakeholders to validate your assumptions.
3. **Include the "do nothing" scenario.** What happens if you don't invest in AI? Often there's a cost to inaction: competitive disadvantage, missed efficiency, continued pain points.
4. **Address objections proactively.** If you know stakeholders have concerns about security, job displacement, or complexity, address them directly before they're raised.
5. **Start small, prove value, expand.** If budget is a concern, propose a pilot that proves the concept with limited investment before requesting full rollout funding.
6. **Tie to strategic priorities.** Connect AI investment to existing company goals. If the company is focused on growth, show how AI enables scaling. If focused on efficiency, emphasize cost reduction.

Key Takeaways

1. **Total cost exceeds subscription price.** Budget 40-60% above software costs for training, integration, and ongoing management. First-year TCO is often nearly double the subscription price.
2. **Subscribe is usually the right choice.** For 90% of SMB AI needs, off-the-shelf subscriptions offer the best balance of cost, speed, and capability. Reserve "build" for genuine competitive differentiation.
3. **Calculate ROI conservatively.** Vendor ROI claims are inflated. Use time savings, capacity, avoided cost, or quality improvement methods, and assume only 25-50% of theoretical savings materialize.
4. **Know when to walk away.** Broken processes, low volume, unmeasurable success, poor data readiness, and organizational unreadiness are signals to invest elsewhere first.
5. **Build stakeholder-specific business cases.** CFOs want ROI, CEOs want strategic alignment, operations wants efficiency. Tailor your case to what each audience values.
6. **Start small, prove value, then scale.** A successful pilot with documented results is the most powerful business case for broader investment.
7. **Account for hidden costs early.** Training, change management, and ongoing administration are real expenses. Ignoring them sets up budget overruns and stakeholder disappointment.

Your 48-Hour Challenge

Before moving to Module 8, complete at least one of these exercises:

1. **Calculate your current AI spend:** List every AI tool your organization uses (including shadow AI). Estimate total annual cost including subscriptions, training time, and management overhead.
2. **Run an ROI calculation:** Pick one AI tool you're using or considering. Apply one of the ROI methods in this module. What does the math say? Does it match your intuition?
3. **Draft a mini business case:** For an AI initiative you're considering, write a one-page summary with problem statement, proposed solution, estimated costs, expected benefits, and key risks.
4. **Evaluate build vs. buy vs. subscribe:** For a specific use case in your organization, work through the decision framework. Which approach makes the most sense and why?

What's Next

The math works. You've run the numbers and the ROI makes sense. So now comes the part that actually determines whether any of this succeeds or fails, and it has nothing to do with technology or spreadsheets.

It's your people. Module 8 is the one I'd read twice if I were you.

THE AI OWNER'S MANUAL

The People Side of AI

Getting your team on board and working through the human challenges of AI adoption

Module Overview

You've built the financial case. You've selected the right tools. You've established governance policies. But there's a force that can stop even the best-planned AI initiative dead in its tracks:

Your People.

AI adoption is fundamentally a change management challenge. The technology may be ready, but your organization is made up of humans with fears, habits, career concerns, and varying levels of enthusiasm for change. Ignore the human element, and your shiny new AI tools will become expensive shelfware. Address it thoughtfully, and you'll build an organization that doesn't just use AI but thrives with it.

This module tackles the people challenges head-on. We'll cover what makes AI change management different from other technology rollouts, how to get buy-in from skeptics, strategies for training and upskilling, the honest conversation about job evolution, how to build internal AI champions, and communication approaches that work.

The goal isn't to manipulate people into accepting AI. It's to address legitimate concerns honestly, provide genuine support for the transition, and create conditions where your team can discover the benefits of AI for themselves.

Learning Objectives

- √ Understand why AI change management requires a different approach than traditional technology rollouts
- √ Apply proven strategies for getting buy-in from AI skeptics without dismissing their concerns
- √ Design effective training and upskilling programs that build genuine competence
- √ Work Through honest conversations about job evolution with transparency and empathy
- √ Build a network of AI champions who accelerate adoption across your organization
- √ Craft communications that inform, inspire, and address concerns proactively

Why AI Change Management Is Different

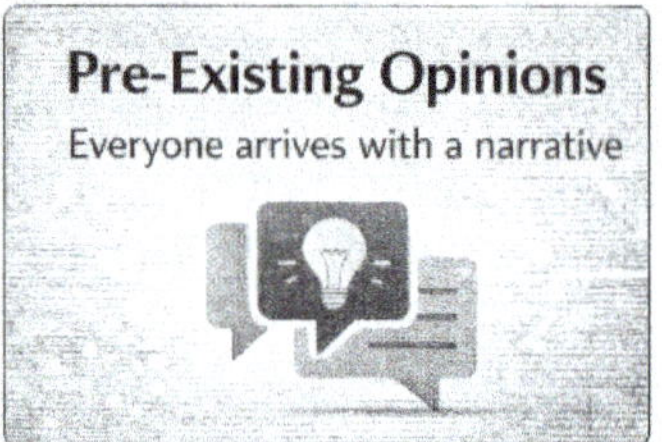

You've probably led technology changes before. New CRM systems, ERP implementations, cloud migrations. AI is different and understanding why helps you anticipate and address challenges that catch many organizations off guard.

AI Touches Identity, Not Just Process

When you implement a new accounting system, people learn new buttons to click. Their role identity stays the same and they're still accountants doing accounting work. AI is different because it can perform tasks that people consider core to their professional identity.

The marketing writer who's been crafting campaigns for twenty years isn't just learning a new tool, they're watching a machine produce passable first drafts in seconds. The analyst who built a career on Excel mastery sees ChatGPT generate formulas they would have spent hours creating. This isn't just process change; it is an identity threat.

What this means for you: Expect deeper resistance than typical technology changes. People aren't just worried about learning something new, they're questioning their professional value. Your change management approach must address this at the identity level, not just the skill level.

The Pace Is Unprecedented

Traditional enterprise software changes slowly. You implement a system, use it for years, maybe get an upgrade every few years. AI capabilities are advancing monthly. The tool you trained people on last quarter might have significant new features (or competitors) today.

This creates unique challenges. People barely feel competent before something changes. Best practices haven't solidified. Even experts are learning in real-time.

What this means for you: Build adaptability, not just proficiency. Training programs must teach learning-to-learn skills, not just current-tool mastery. Create culture where continuous adjustment is expected and supported.

The Job Impact Is Real and Visible

With previous technology changes, the "jobs will be affected" narrative was often theoretical or distant. With AI, people can see it happening. They can watch ChatGPT write emails, see Midjourney create images, observe code assistants accelerate development. The impact isn't abstract, it's demonstrated daily in news stories and their own experiments.

This visibility makes reassurance harder. "Your job is safe" rings hollow when someone just watched AI do 30% of their daily tasks in their own browser.

What this means for you: Don't minimize or deny the impact. Honesty about how roles will change, while also providing a clear path through that change, builds more trust than false reassurance.

Everyone Has an Opinion

When you implemented that new CRM, most employees didn't have strong preconceived notions. AI is different. Everyone has been exposed to AI narratives, from apocalyptic predictions to utopian promises, through news, social media, and entertainment. They arrive at your AI initiative with opinions already formed.

Some are excited and expect magic. Others are terrified based on dystopian scenarios. Many are skeptical after disappointing experiences with earlier AI hype cycles. You're not introducing AI to blank slates. You're introducing it to people with existing mental models, many of which are inaccurate.

What this means for you: Start by understanding what people already believe. Your first job may be correcting misconceptions before you can build an accurate understanding.

Getting Buy-In from Skeptics

AI skeptics aren't your enemy. They're often your most valuable allies in disguise. Their concerns frequently point to real issues that enthusiasts overlook. Your goal isn't to defeat skeptics but to understand their concerns, address legitimate ones, and help them discover value on their own terms.

Understanding the Types of Skeptics

Not all skepticism is the same, and different concerns require different responses.

The Quality Skeptic

What they say: "AI output isn't good enough for our standards. It makes errors. It can't understand nuance."

What they mean: "I've worked hard to develop expertise, and I'm not willing to let a machine produce inferior work under my name."

How to respond: These skeptics are often your best quality gatekeepers. Acknowledge their standards as valuable. Position AI as a first-draft tool that still requires their expertise. Show them how AI can handle routine work, freeing them for the nuanced tasks where their expertise matters most. Let them evaluate AI output critically, their feedback improves how everyone uses it.

The Job Security Skeptic

What they say: "We're just training our replacements."

What they mean: "I'm afraid this technology will make me irrelevant, and I'm being asked to accelerate that process."

How to respond: This fear deserves honest acknowledgment, not dismissal. Be direct about how you see roles evolving (we'll cover this in detail later in the module). If you genuinely don't plan to eliminate positions, say so clearly and explain why. If some roles will change significantly, acknowledge that while showing the path forward. The worst response is vague reassurance that doesn't address the actual concern.

The Burned-Before Skeptic

What they say: "We've seen this before. Remember the chatbot project? The automation initiative? This is just another overhyped technology."

What they mean: "I've invested time and energy in previous initiatives that didn't deliver. Why should this be different?"

How to respond: Acknowledge past disappointments explicitly. Explain what's different now. The technology has genuinely improved, not just the marketing. Start with small, demonstrable wins that prove value before asking for larger commitments. Let them be skeptical while still participating. Their critical eye helps keep expectations realistic.

The Ethics Skeptic

What they say: "What about bias? What about privacy? What about the environmental impact? Are we sure this is responsible?"

What they mean: "I care about doing the right thing, and I'm not convinced this technology is ethical."

How to respond: These concerns are legitimate and should inform your governance approach (Module 6). Invite ethics skeptics into policy discussions. Their perspective makes your AI governance stronger. Show the guardrails you're putting in place. Be honest about limitations and uncertainties. Their vigilance helps the organization use AI responsibly.

The Skeptic Conversion Playbook

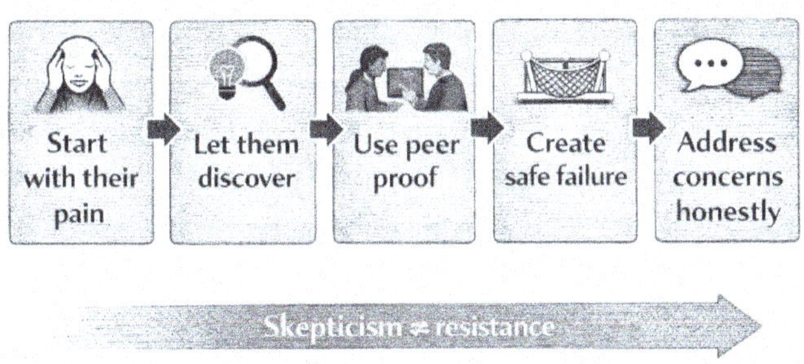

Beyond understanding different skeptic types, here are proven strategies for building buy-in.

Start with Their Pain Points

Don't lead with AI capabilities. Lead with problems they already want solved. "I know you spend hours on monthly reports. Want to see something that might help?" is more effective than "Let me show you this amazing AI tool." When AI solves a problem they care about, adoption follows naturally.

Let Them Discover, Not Be Told

Mandates create resistance; discovery creates enthusiasm. Rather than telling skeptics how great AI is, give them opportunities to experiment in low-pressure environments. Provide access, offer suggestions, but let them find value on their own terms. Personal discovery is more persuasive than any presentation.

Use Social Proof Wisely

Skeptics often respond better to peers than to leadership. When someone on their level, especially someone they respect, starts getting results with AI, curiosity follows.

Identify respected team members who are open to AI, support their experimentation, and let their results speak for themselves.

Create Safe Failure Spaces

Fear of looking foolish prevents experimentation. Create environments where trying AI and not getting good results is explicitly acceptable. "Sandbox" time for experimentation, learning sessions where mistakes are shared and learned from, and leadership modeling their own AI learning curve all help reduce fear.

Address Concerns, Don't Dismiss Them

Never respond to concerns with "you just don't understand" or "you'll come around." Every concern deserves a serious response. If you can't address a concern, acknowledge it honestly: "That's a real limitation right now. Here's how we're managing it, and here's what we're watching for."

Training and Upskilling Strategies

Most AI training programs fail because they teach the wrong things in the wrong ways. They focus on tool features rather than practical workflows. They're one-time events rather than ongoing development. They assume everyone starts at the same place. Here's how to do it better.

The AI Training Hierarchy

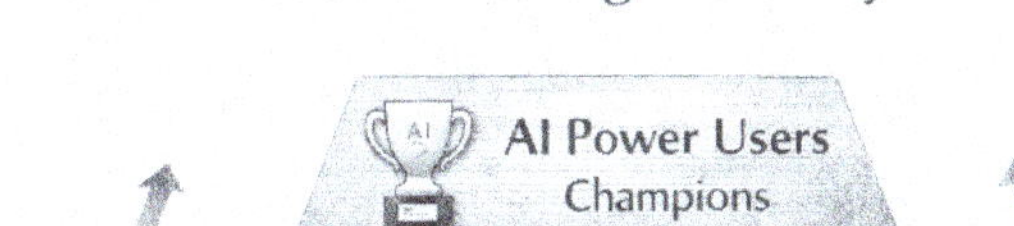

Effective AI training builds competence in layers, starting with foundations and progressing to advanced applications.

Level 1: AI Literacy

Goal: Everyone understands what AI is, what it can and can't do, and how to use it safely.

Content: Basic AI concepts in plain language, company AI policies and acceptable use, simple use cases relevant to their role, and security and privacy guidelines.

Format: 60-90 minute session, in-person or video, with Q&A.

Who: Everyone in the organization.

Level 2: AI User Competence

Goal: Team members can effectively use AI tools for common tasks in their role.

Content: Specific tools used in your organization, role-specific use cases and workflows, prompt engineering fundamentals (Module 5), practical exercises with real work examples.

Format: 2-4 hours, hands-on workshops, spread over multiple sessions.

Who: Knowledge workers who will use AI regularly.

Level 3: AI Power User

Goal: Selected individuals become highly proficient and can help others.

Content: Advanced prompt techniques, workflow integration, building custom applications, troubleshooting common issues, staying current with new capabilities.

Format: Ongoing development, project-based learning, peer collaboration.

Who: AI champions, team leads, those with aptitude and interest.

Training Design Principles

Use Their Actual Work

Generic examples produce generic learning. Training should use real documents, real emails, real tasks from participants' actual work. "Let's write a product description" is less effective than "Let's use AI to help with that proposal you're working on." When people see AI working on their problems, learning accelerates.

Teach Judgment, Not Just Mechanics

Knowing how to use AI is less important than knowing when to use it and when not to. Include decision-making training: When is AI the right tool? How do you evaluate whether output is good enough? What requires human judgment even with AI assistance? Critical thinking about AI is as important as operational skill.

Make It Continuous

One-time training events decay quickly, especially with rapidly evolving tools. Build ongoing mechanisms: monthly "tips and tricks" sessions, shared channels for questions and discoveries, regular capability updates as tools evolve, peer learning communities. Treat AI competence as continuous development, not one-time certification.

Meet People Where They Are

A 25-year-old who's been experimenting with ChatGPT for two years needs different training than a 60-year-old who's never used generative AI. Use assessments to understand current skill levels. Offer different tracks or pacing. Don't bore advanced users or overwhelm beginners by forcing everyone through identical content.

Sample Training Program Structure

Here's a practical framework for rolling out AI training across your organization.

Week	Focus	Activity	Outcome
1	AI Literacy	All-hands session: What AI is, isn't, and our approach	Shared understanding, reduced anxiety
2-3	Hands-On Basics	Small group workshops by department	Basic competence with approved tools
4-6	Role-Specific	Workflows for specific job functions	Practical application to daily work
7-8	Champion Development	Advanced training for selected power users	Internal support network
Ongoing	Continuous Learning	Monthly sessions, shared channels, office hours	Sustained competence and improvement

The Honest Conversation: Job Evolution vs. Elimination

This is the elephant in every AI room. People want to know: "Is AI going to take my job?" They deserve an honest answer, even when that answer is complicated.

Job Evolution vs Job Elimination

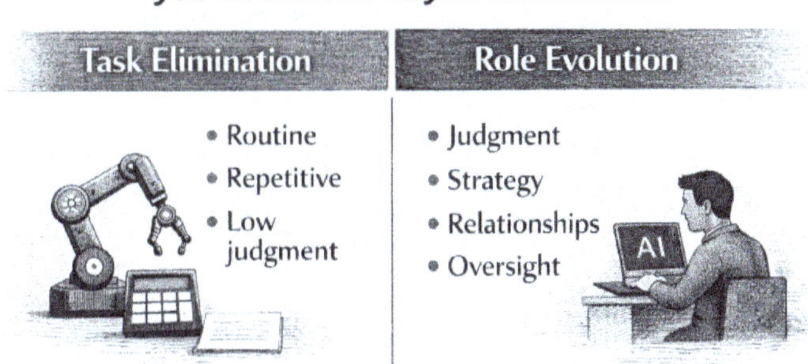

What the Research Actually Shows

Despite apocalyptic headlines, the reality is more nuanced. Most serious research suggests AI will transform jobs more than eliminate them, at least in the medium term.

Tasks, not entire jobs, are what AI typically automates. Few jobs consist entirely of automatable tasks.

The pattern that's emerging is task redistribution. AI handles routine, repetitive, and data-processing tasks. Humans shift to tasks requiring judgment, creativity, relationship-building, and complex problem-solving. Total employment may not change dramatically, but the nature of many jobs will.

That said, some roles will genuinely shrink. Data entry, basic content production, simple customer service queries, routine document processing all face significant automation pressure. Organizations need to be honest about this while providing transition support.

How to Have the Conversation

When discussing AI's impact on jobs with your team, here's how to approach it with integrity.

Be Honest About Uncertainty

Nobody knows exactly how AI will reshape work over the next decade. Saying "nothing will change" is as dishonest as saying "everything will change." Acknowledge uncertainty while sharing what you can control: your organization's values, your commitment to supporting employees through transition, and your plans for the near term.

Distinguish Your Organization from Headlines

People read about mass layoffs at tech companies attributed to AI. They need to understand how your organization is different or similar. If you're a 50-person professional services firm, your situation isn't identical to a tech company laying off thousands. Be specific about your business model, your growth plans, and how AI fits into them.

Focus on What You Can Promise

You may not be able to guarantee no job changes ever. But you can promise investment in training, advance notice of significant changes, support for transitions, and prioritizing internal mobility over layoffs. Make commitments you can keep and be explicit about what you're committing to.

Reframe from Threat to Opportunity

The honest message isn't "AI won't change your job" (which may not be true). It's "Your job will evolve, and we're investing in helping you evolve with it." Position AI competence as career insurance. People who learn to work effectively with AI are more valuable, not less.

The Evolution Framework

Help people understand how their specific roles might evolve by walking through this framework.

Task Type	AI's Role	Human's Evolved Role
Routine Data Processing	AI handles most processing	Human designs processes, handles exceptions, ensures quality
Content Creation	AI generates first drafts, variations, routine content	Human provides strategy, brand voice, editing, quality control
Analysis	AI processes data, identifies patterns, generates initial insights	Human interprets meaning, applies context, makes decisions
Customer Interaction	AI handles routine inquiries, prepares information, drafts responses	Human handles complex situations, builds relationships, exercises judgment
Strategy & Planning	AI provides research, scenarios, competitive intelligence	Human sets direction, makes trade-offs, leads execution

Have individuals map their own roles through this framework. Which of their tasks fall into each category? How might their role shift if routine tasks are automated? This exercise makes abstract change concrete and personal.

Job Evolution vs Job Elimination

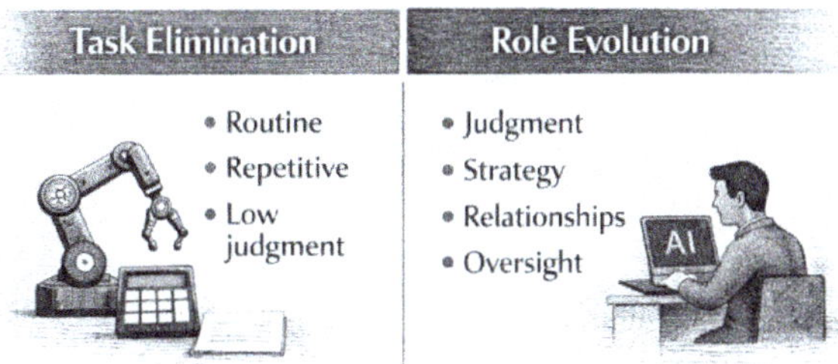

Building AI Champions

AI champions are employees who accelerate adoption throughout your organization. They're not just users. They're advocates, teachers, and problem-solvers who help others discover AI's value. A strong champion network can be the difference between successful adoption and expensive failure.

What Makes a Good AI Champion

Not everyone with AI enthusiasm makes a good champion. Look for this combination of traits.

- **Credibility in their domain.** Champions need respect from peers. Someone seen as technically competent and good at their job carries more weight than someone seen as just interested in new technology.
- **Genuine enthusiasm, not hype.** Look for people who are excited but realistic and who acknowledge limitations while seeing possibilities. Overhypers can damage credibility.
- **Patience and teaching ability.** Champions will field basic questions repeatedly. They need patience to help people at all skill levels without condescension.
- **Cross-functional relationships.** The best champions know people across the organization. They can spread knowledge beyond their immediate team.
- **Willingness to experiment and share.** Champions try new things, document what works, and share their discoveries openly including failures.

The Champion Program Structure

Formalize your champion network with clear roles and support.

Selection

Identify 1-2 champions per department or team of 10-15 people. Some may self-nominate; others should be recruited based on the traits above. Make it an opt-in role as forced champions rarely succeed.

Training

Champions receive advanced training (Level 3 from our training hierarchy). They should be comfortable with all approved tools, understand company AI policies deeply, and have exposure to advanced techniques. Training should also cover how to teach and support others effectively.

Responsibilities

Clearly define what champions are expected to do:

- Be the first point of contact for AI questions in their area
- Run informal training sessions or lunch-and-learns
- Share successful use cases and workflows
- Gather feedback and suggestions from their teams
- Participate in the champion community and share learnings
- Stay current on new capabilities and policy updates

Support

Champions need time and resources. Allocate 2-4 hours per week for champion activities. Provide early access to new tools and features. Create a champion community where they can share challenges and solutions. Give them direct access to leadership for escalating issues.

Recognition

Make champion contributions visible. Include champion work in performance discussions. Recognize successful adoption in their areas. Consider the champion role as leadership development. People invest more in roles that advance their careers.

Communication Templates and Strategies

How you communicate about AI significantly impacts how people receive it. Poor communication breeds rumors, fear, and resistance. Thoughtful communication builds understanding, reduces anxiety, and creates conditions for successful adoption.

Communication Principles

Lead with "Why"

Don't start with the technology, start with the business reason. "We're implementing AI tools because..." should be followed by business goals people understand: serving customers better, reducing tedious work, staying competitive, growing the business. Technology for its own sake generates skepticism.

Acknowledge Concerns Proactively

Don't wait for concerns to surface as rumors. Address the obvious questions upfront: How will this affect jobs? What about privacy? Will I be required to use this? Proactive acknowledgment shows you understand what people are thinking and builds trust.

Be Specific About What's Changing

Vague announcements create anxiety. "We're bringing AI to the organization" raises more questions than it answers. "We're providing access to Claude for our marketing team, starting with content editing and brainstorming" is concrete and manageable.

Create Feedback Channels

One-way communication feels like pronouncement. Two-way communication feels like partnership. Provide clear ways for people to ask questions, raise concerns, and share feedback. And then actually respond to what you hear.

Sample Communications

Below are templates you can adapt for your organization. Customize the specifics while keeping the structural approach.

Template 1: Initial AI Initiative Announcement

Subject: Introducing AI Tools to Help Us Work Smarter

Team,

I'm writing to share some news about how we're evolving our toolkit to serve our clients better and reduce the time you spend on routine tasks.

Starting [date], we'll be introducing AI-assisted tools to help with [specific use cases]. This is part of our commitment to [business objective that resonates with the team].

I know you probably have questions. Let me address a few upfront:

Why are we doing this? [Clear business rationale] / Will this affect my job? [Honest, specific answer] / What's expected of me? [Clear expectations and timeline]

We're committed to supporting you through this transition with training, resources, and open communication. [Name] will be leading our AI champion program and is available to answer questions.

There will be a Q&A session on [date/time]. In the meantime, feel free to reach out to [contact] with questions.

Template 2: Training Session Invitation
Subject: Your AI Training Session – [Date]
Hi [Name],

You're invited to our AI tools training session on [date] at [time]. This [duration] session will give you hands-on experience with the tools we're introducing.

What you'll learn: [2-3 specific, relevant outcomes]

What to bring: A work project or task you'd like to try with AI assistance

No prior AI experience needed. We'll start with the basics and work up to practical applications for your role.

[Register/confirm attendance link]

Template 3: Progress Update
Subject: AI Initiative Update – Month [#]
Team,

Here's where we stand with our AI tools after [time period]:

What's working: [Specific wins with metrics if available]

What we're improving: [Challenges being addressed]

Coming next: [Upcoming changes or expansions]

Spotlight: [Highlight a success story or team member making good use of AI]

Questions or suggestions? Reach out to [contact] or join our next Q&A on [date].

Key Takeaways

1. **AI change management is different.** It touches identity, moves faster than traditional technology change, has visible job impacts, and arrives with preconceived notions. Standard change management approaches need adaptation.
2. **Skeptics are valuable allies.** Different skeptic types have different concerns: quality, job security, past disappointments, ethics. Address concerns honestly rather than dismissing them, and let skeptics discover value on their own terms.
3. **Training must be continuous and practical.** Build competence in layers: literacy, user competence, power user. Use real work examples, teach judgment alongside mechanics, and make learning ongoing rather than one-time.
4. **Have honest job conversations.** AI will change roles but mostly through task redistribution rather than elimination. Be honest about uncertainty while committing to training, support, and transparent communication about changes.
5. **Champions accelerate adoption.** Identify credible enthusiasts, give them advanced training and time allocation, define clear responsibilities, and recognize their contributions. A strong champion network can be the difference between success and failure.
6. **Communication makes or breaks adoption.** Lead with business reasons, acknowledge concerns proactively, be specific about what's changing, and create genuine two-way feedback channels. How you communicate matters as much as what you communicate.

Your 48-Hour Challenge

Before moving to Module 9, complete at least one of these exercises:

1. **Map your skeptics:** Identify the main skeptics in your organization. What type of skeptic is each one? What specific concerns might they have? Draft a personalized approach for engaging with each.

2. **Design your training program:** Using the hierarchy and timeline in this module, sketch out a training program for your organization. Who gets what level? What's the realistic timeline? What resources do you need?

3. **Identify potential champions:** Make a list of 3-5 people who could be effective AI champions based on the criteria in this module. What would you need to do to recruit and support them?

4. **Draft your announcement:** Using the templates as a starting point, write the initial AI announcement you would send to your team. Have someone review it for tone and clarity.

What's Next

Getting your team on board is necessary. But here's the thing: your competitors are doing the same training, having the same conversations, overcoming the same resistance. Everyone is going to get their people comfortable with AI eventually.

The question that matters for the long term is different. Module 9 asks: once everyone has AI, what makes *you* hard to compete with?

THE AI OWNER'S MANUAL

Building Your AI Advantage

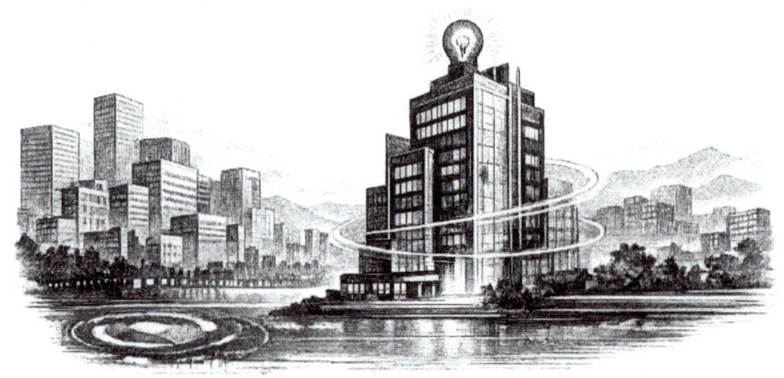

Moving from AI as a productivity tool to AI as your business's competitive moat.

Module Overview

You've built the foundation. You've implemented AI tools, trained your team, established governance, and started seeing results. But here's the uncomfortable truth that most AI education ignores:

Efficiency gains disappear the moment your competitors adopt the same tools.

The AI tools you're using today are available to everyone. ChatGPT, Claude, and Copilot are tools your competitors can sign up for tomorrow and achieve the same productivity gains you've worked hard to implement. In a world where everyone has access to the same AI capabilities, where does competitive advantage come from?

This module tackles the strategic question that separates companies that *use* AI from companies that *compete* with AI.

We'll explore where AI creates defensible, lasting advantages versus temporary efficiency gains. We'll examine how your data can become a proprietary asset that makes your AI capabilities impossible to replicate. And we'll build a strategic roadmap that compounds value over time rather than just keeping pace with competitors.

The goal isn't to chase every AI advancement. It's to make strategic choices about where AI can create genuine differentiation for your specific business. Then give your business the knowledge to build capabilities that become harder to replicate the longer you invest in them.

Learning Objectives

- ✓ Distinguish between AI capabilities that create temporary efficiency versus lasting competitive advantage
- ✓ Identify opportunities to build proprietary data assets that strengthen your AI capabilities over time
- ✓ Apply frameworks for AI-driven customer experience differentiation
- ✓ Evaluate industry-specific AI opportunities with realistic competitive analysis
- ✓ Develop a 1-year and 3-year AI strategic roadmap aligned with business objectives
- ✓ Build organizational differentiation

The Efficiency Trap: Why Speed Isn't Strategy

Most companies approach AI with a simple logic: AI makes things faster, faster is better, therefore AI is good. This thinking leads to what we call the *efficiency trap*, investing in AI improvements that your competitors can easily match and leaving you running faster on the same treadmill.

The Competitive Dynamics of AI Adoption

Let's consider what happens when you implement an AI tool that cuts your content creation time in half. You just gained an advantage. You can now produce more content at a lower cost than your competitors and at a rate that they're struggling to match. But within months, your competitors adopt the same or similar tools. Now everyone creates content at the new speed. The efficiency gain has become table stakes, not differentiation.

This pattern repeats across nearly every generic AI application: email assistance, document drafting, basic analysis, scheduling optimization. These tools create real value, your organization *should* use them, but they don't create competitive advantage because they're equally accessible to everyone.

The critical question isn't "How can AI make us faster?" It's "How can AI make us **different** in ways that are hard to copy?"

The Four Types of AI Value

Not all AI implementations are equal in their strategic value. Understanding where your AI investments fall helps prioritize resources and set realistic expectations.

Type	Description	Example	Strategic Value
Commodity Efficiency	Generic tools anyone can adopt	Email drafting, basic content generation	Low - Table stakes already today
Process Advantage	AI integrated into unique workflows	Custom quoting systems, specialized QC automation	Medium - Replicable but requires effort
Data Advantage	AI trained on proprietary data	Prediction models using historical company data	High - Data can't be easily replicated
Network Advantage	AI improves as more users/data flow through	Customer behavior models, supplier matching systems	Very High - Creates compounding moat

Your strategy should include Commodity Efficiency investments as they're necessary to stay competitive but your **strategic focus** should be on building Data Advantage and Network Advantage capabilities where possible.

Remember, Data isn't just king, it's your most valuable asset.

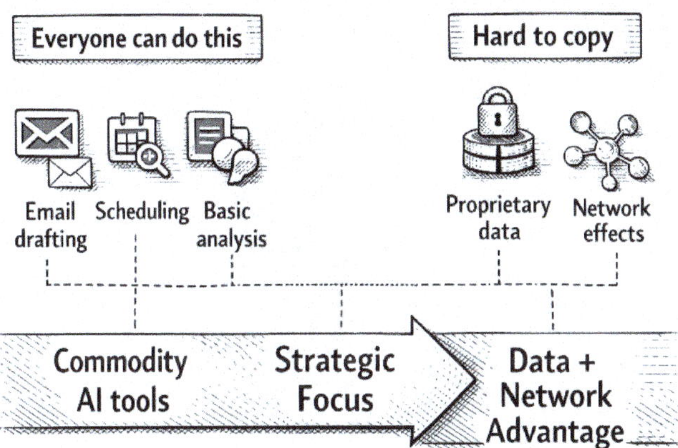

Data as Your AI Asset

The companies building real AI competitive advantage understand a fundamental truth: the AI models themselves are increasingly commoditized, but **the data that makes them useful for your specific context is not.** Your historical data, your customer interactions, your operational patterns. These are assets that competitors simply cannot replicate.

The Data Flywheel

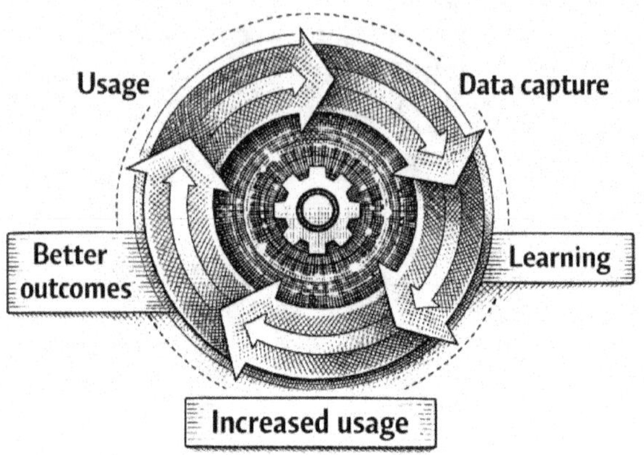

The most powerful AI advantages come from data flywheels. Systems where using AI generates data that makes the AI better, which generates value, which inevitably generates more data. This creates a compounding advantage that widens over time.

Example: A distribution company's pricing advantage.

Consider a regional distributor who implements AI-assisted pricing. Initially, the AI provides modest improvements over manual pricing. But every pricing decision (and the customer response to that decision) becomes training data. Over 18 months, the system has learned which customers are price-sensitive on which products, which combinations sell together, and which seasonal patterns affect specific product categories. A competitor starting fresh with the same AI technology can't replicate 18 months of accumulated, context-specific learning.

Identifying Your Data Assets

Most SMBs significantly underestimate the data assets they already possess. The key is recognizing data that is unique to your organization, accumulates over time, directly relates to value-creating activities, and is difficult or impossible for competitors to obtain.

Data Category	Examples	AI Applications
Customer Interaction History	Support tickets, sales conversations, feedback, complaints, purchase patterns	Churn prediction, personalized recommendations, proactive service
Operational Data	Process timing, quality outcomes, equipment performance, resource utilization	Predictive maintenance, capacity optimization, quality prediction
Expert Knowledge	How experienced employees solve problems, tribal knowledge, decision patterns	AI assistants that embody company expertise, training acceleration
Market-Specific Data	Local market conditions, regional patterns, customer segment behaviors	Localized predictions, segment-specific optimization

Building Your Data Strategy

Converting existing data into AI advantage requires deliberate effort. Here's a practical framework for building your data strategy:

Step 1: Inventory existing data. Document what data you currently collect, where it lives, and how accessible it is. Most organizations discover they have more usable data than they realized. It's just scattered and unstructured.

Step 2: Identify capture gaps. What valuable interactions or decisions happen without being recorded? The expert who resolves complex issues, the sales rep who knows which pitch works for which customer type, and the operations manager who can predict problems before they happen all hold knowledge that often exists only in their heads.

Step 3: Design capture mechanisms. Build lightweight processes to capture valuable data. This doesn't require expensive systems as sometimes it's as simple as structured templates for recording decisions, regular knowledge capture sessions, or AI tools that log interactions they assist with.

Step 4: Create feedback loops. The most valuable data captures not just decisions but outcomes. Did the AI-assisted recommendation work? Did the predicted issue actually occur? Feedback data is what enables AI systems to learn and improve.

Step 5: Protect and govern. Your data assets need protection. This means both security measures and governance policies that ensure data quality and appropriate use. Poor data hygiene today creates AI problems tomorrow.

Customer Experience Differentiation

One of the most powerful applications of AI for competitive advantage is customer experience. Unlike internal efficiency gains that are invisible to the market, customer experience improvements create differentiation that customers directly perceive and value.

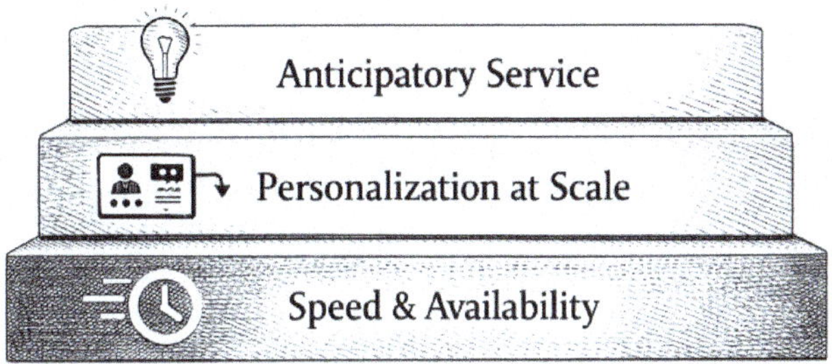

The Three Layers of AI-Enhanced Experience

Layer 1: Response Speed and Availability. The most basic application. AI that responds quickly and never sleeps. This includes chatbots, automated responses, and 24/7 availability. Important to implement, but increasingly expected rather than differentiating.

Layer 2: Personalization at Scale. AI that remembers customer context and tailors interactions accordingly. This moves beyond "Hello [First Name]" to genuinely customized recommendations, proactive suggestions based on history, and communications that reflect the specific customer relationship.

Layer 3: Anticipatory Service. The highest level. AI that predicts customer needs before they're expressed. Reaching out before problems occur, suggesting solutions to issues customers haven't yet recognized, and creating value customers didn't know to ask for.

Most companies focus on Layer 1 because it's easiest. The competitive advantage lies in Layers 2 and 3 and these require the proprietary data assets we discussed earlier.

Practical Applications by Business Type

Professional Services: AI that reviews client history before calls, drafts personalized proposals based on past work, identifies cross-selling opportunities based on similar client patterns, and proactively alerts account managers to potential issues.

Distribution/Wholesale: AI that predicts when customers need to reorder, suggests complementary products based on buying patterns, alerts sales to unusual order gaps, and provides customers with usage insights they can't get elsewhere.

Manufacturing: AI that provides real-time order status with predictive accuracy, proactively communicates potential delays, suggests alternative configurations based on production capabilities, and learns customer quality preferences.

Retail/E-commerce: AI that creates genuinely personalized shopping experiences, predicts returns before they happen, identifies loyal customers at risk of churning, and customizes communication timing and channels.

The Human-AI Balance

AI Handles	Humans Handle	AI Enhances Humans
• Routine tasks	• Empathy	• Decision support
• Data crunching	• Ethical decisions	• Pattern recognition
• Basic analysis	• Creative problem-solving	• Task automation
Speed & Availability		

A critical strategic decision is where AI augments human interaction versus replaces it. The answer isn't universal. It depends on your customers, your value proposition, and where human touch genuinely adds value.

AI should handle: routine inquiries, information lookup, initial triage, scheduling, standard transactions, and monitoring for issues.

Humans should handle: complex problem-solving, emotional situations, relationship building, strategic discussions, and exceptions that require judgment.

AI should enhance humans: providing context and history before interactions, suggesting next best actions, handling post-call documentation, and identifying patterns humans might miss.

The winning formula for most SMBs isn't maximum automation. It's strategic automation that frees humans to do what they do best while AI handles the rest.

Operational Excellence at Scale

Internal operations might not directly face customers, but AI-driven operational excellence creates competitive advantage through lower costs, faster delivery, better quality, and more consistent execution.

From Point Solutions to Integrated Intelligence

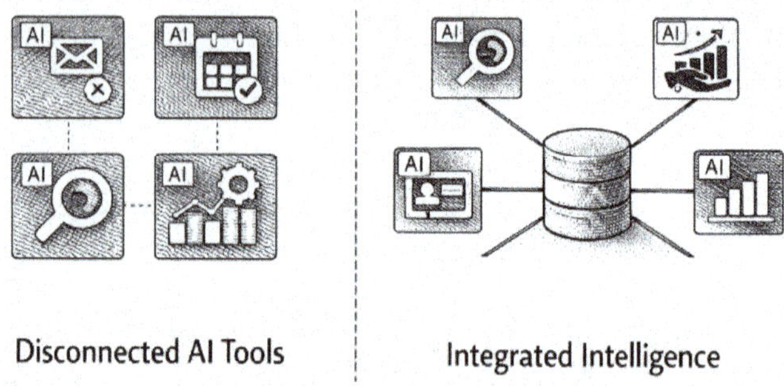

Most companies start with point solutions. An AI tool for this process, another for that task. This is appropriate for early stages. But lasting operational advantage comes from integration, where AI capabilities connect across functions and share learning.

Example: Connected operations in a manufacturing context. A manufacturer starts with separate AI tools: one for demand forecasting, another for production scheduling, a third for quality control. Each provides modest improvements. But when these systems share data and learning, when quality data informs scheduling, scheduling informs procurement, and all of it informs forecasting, the combined system becomes significantly more powerful than the sum of its parts. Problems identified in quality automatically adjust scheduling. Demand patterns automatically flow through to supplier orders. The system learns as a whole.

This integration takes time and deliberate architecture. It's not where you start, but it's where significant operational advantage emerges.

Decision Automation Spectrum

Operational AI exists on a spectrum from pure human decision-making to full automation. Understanding where different decisions should fall helps you extract maximum value while managing risk.

Level	Description	Appropriate For	Example
AI Informs	AI provides data/insights; human decides	High-stakes, complex, or judgment-intensive decisions	Strategic pricing, major customer negotiations
AI Recommends	AI suggests action; human approves	Important decisions with clear patterns and some exceptions	Inventory reorders, production scheduling
AI Acts, Human Reviews	AI executes; human spot-checks	Frequent, lower-stakes decisions with good patterns	Routine customer inquiries, standard reports
Full Automation	AI handles entirely	High volume, clear rules, low individual stakes	Data entry validation, standard categorization

Start most decisions at lower automation levels and move up as confidence increases. It's much easier to increase automation than to recover from errors caused by premature full automation.

Industry-Specific Opportunities

While AI principles apply across industries, the specific opportunities for competitive advantage vary significantly. Here's a realistic assessment of where AI creates the greatest strategic value in different sectors.

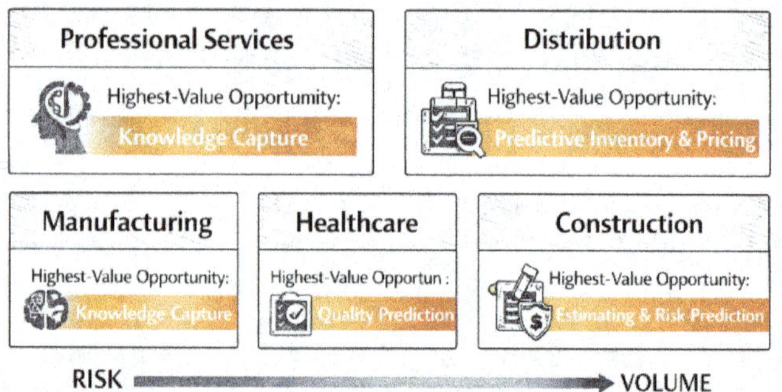

Professional Services

Highest-value opportunity: Knowledge capture and leverage. Professional services firms sell expertise, but that expertise often walks out the door each night and eventually retires. AI can capture, codify, and make accessible the accumulated wisdom of your best practitioners.

Specific applications: AI assistants trained on firm knowledge, automated first drafts of common deliverables, intelligent research and precedent search, proposal generation based on past successful pitches.

Competitive consideration: The firms that capture and systematize their expertise first will compound that advantage. Waiting means competitors build larger knowledge bases.

Distribution and Wholesale

Highest-value opportunity: Predictive inventory and dynamic pricing. Distribution margins are thin, and the difference between good and great inventory management directly impacts profitability.

Specific applications: Demand forecasting that accounts for local patterns, automated reorder optimization, dynamic pricing based on inventory position and customer segments, predictive identification of customer needs.

Competitive consideration: Historical transaction data is your moat. Competitors can adopt the same AI tools, but they can't replicate your customer relationship data.

Manufacturing

Highest-value opportunity: Quality prediction and process optimization. Every manufacturer knows quality matters, but most catch defects after they occur. AI can predict quality issues before they happen.

Specific applications: Predictive quality based on process parameters, predictive maintenance to prevent downtime, production scheduling optimization, energy use optimization.

Competitive consideration: Manufacturing AI advantages compound with time. The longer you've been collecting process data, the better your predictions become.

Healthcare and Life Sciences

Highest-value opportunity: Documentation and administrative automation. Clinical staff spend enormous time on documentation; AI can reduce this burden significantly while improving consistency.

Specific applications: Clinical documentation assistance, prior authorization automation, patient communication, appointment optimization, research literature monitoring.

Competitive consideration: Regulatory requirements (HIPAA, etc.) constrain AI options. Organizations that figure out compliant AI deployment first gain significant advantage.

Construction and Trades

Highest-value opportunity: Estimating and project risk prediction. Accurate estimates win jobs; poor estimates lose money. AI can significantly improve estimation accuracy using historical project data.

Specific applications: AI-enhanced estimating using historical projects, change order prediction, scheduling optimization, safety incident prediction, subcontractor performance analysis.

Competitive consideration: Years of project data create an insurmountable advantage for AI-powered estimation. Start capturing and structuring this data now.

Building Your Strategic AI Roadmap

Moving from AI experimentation to AI strategy requires a roadmap that balances quick wins with long-term capability building. This section provides frameworks for developing 1-year and 3-year plans.

The Phased Approach

AI strategy unfolds in phases. Trying to jump directly to advanced capabilities without building foundations leads to expensive failures.

Phase 1: Foundation (Months 1-6). Focus on building baseline capabilities. Basic AI literacy across the organization, governance frameworks, initial tool deployments, and data inventory. Success in this phase is about establishing competence and trust, not transformation.

Phase 2: Expansion (Months 7-18). Expand successful pilots, integrate AI into more processes, begin building proprietary data assets, and start connecting point solutions. Success in this phase is measured by adoption depth and initial ROI demonstration.

Phase 3: Differentiation (Months 19-36). This is where competitive advantage emerges. Data flywheels begin to spin, integrated systems create unique capabilities, and customer-facing AI differentiates your offering. Success is measured by capabilities that competitors cannot easily replicate.

The 1-Year Roadmap Framework

A realistic 1-year roadmap for an SMB moves from foundation through early expansion:

Quarter	Focus Areas	Success Metrics
Q1: Foundation	Complete AI literacy training, establish governance framework, identify 3-5 pilot opportunities, conduct data inventory	80%+ leadership AI-literate, governance policy approved, pilot projects selected
Q2: First Pilots	Launch 2-3 pilots, begin tracking metrics, identify champions, start team training	Pilots active with defined metrics, champion network established, 50%+ of target users trained
Q3: Evaluate and Expand	Assess pilot results, expand successful deployments, begin data capture improvements, identify integration opportunities	Documented ROI from pilots, expanded deployments launched, data capture plan in progress
Q4: Scale and Plan	Scale successful initiatives, calculate full-year ROI, develop Year 2 strategy, identify strategic AI opportunities	Measurable business impact, Year 2 plan approved, strategic opportunities identified

The 3-Year Strategic Vision

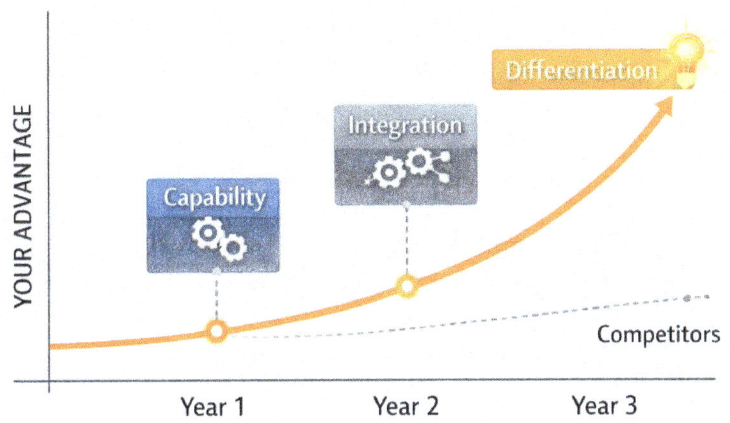

Looking further ahead helps ensure today's decisions build toward meaningful competitive advantage:

Year 1: Establish foundation, prove value, build organizational capability. Most AI initiatives should show measurable ROI by year-end.

Year 2: Scale proven applications, begin integrating systems, focus on building data assets. This is typically when AI moves from "initiative" to "how we work."

Year 3: Differentiation should emerge. Proprietary capabilities, customer-facing AI advantages, and integrated operational systems should create value competitors can't easily match.

Strategic Questions for Your Roadmap

As you develop your roadmap, consider these strategic questions:

1. **Where does AI align with existing strategic priorities?** AI should accelerate your strategy, not become a strategy of its own.
2. **What data assets could become competitive advantages?** Identify data that is unique, accumulating, and related to value creation.
3. **Where does AI touch the customer experience?** Customer-facing applications create visible differentiation.
4. **What organizational capabilities need development?** People, processes, and culture enable AI success.
5. **How will you measure success beyond efficiency metrics?** Strategic value requires strategic measures.
6. **What is your competitive response if competitors move faster?** Have a plan for different competitive scenarios.

Avoiding Strategic Pitfalls

As companies move from AI experimentation to AI strategy, several common mistakes derail progress.

The "AI for AI's Sake" Trap

Some organizations pursue AI because it's the hot technology, not because it solves real business problems. They implement AI tools, create AI initiatives, and hire AI resources, but the connection to business outcomes remains vague. Strategy requires discipline. Every AI investment should have a clear line to business value.

The "Boil the Ocean" Mistake

Ambitious roadmaps that try to transform everything simultaneously almost always fail. Resources spread thin, focus dissipates, and nothing reaches maturity. Better to achieve genuine advantage in two or three areas than superficial capability in twenty.

The Data Quality Assumption

Many AI strategies assume data is ready to use. Reality is messier. Data is scattered, inconsistent, incomplete, or formatted incorrectly. Building AI advantage requires honest assessment of data readiness and investment in data quality as a foundation.

The "Set and Forget" Mentality

AI systems require ongoing attention. Models drift, data patterns change, and capabilities evolve. Organizations that deploy AI and move on find their advantages eroding. Plan for continuous improvement, monitoring, and updating.

Ignoring the Organizational Challenge

Strategic AI plans that focus only on technology and ignore culture, skills, and change management fail to realize their potential. Capability building must be part of the strategy.

Key Takeaways

1. **Efficiency alone isn't strategy.** Generic AI tools create temporary advantages that competitors quickly match. Focus strategic attention on building AI capabilities that are hard to replicate, especially those involving proprietary data.
2. **Data is your AI moat.** AI models are increasingly commoditized, but your data is not. Invest in capturing, structuring, and leveraging data that is unique to your organization. This creates a compounding advantage over time.
3. **Customer experience creates visible differentiation.** Internal efficiency improvements are valuable but invisible. AI-enhanced customer experiences create differentiation that customers directly perceive and value.
4. **Integration multiplies value.** Point solutions provide point improvements. Connected AI systems that share learning create capabilities greater than the sum of their parts.
5. **Industry context shapes opportunity.** The highest-value AI applications vary by industry. Understand where AI creates the greatest advantage in your specific sector and focus resources there.
6. **Strategic roadmaps build in phases.** Foundation, then expansion, then differentiation. Trying to skip phases leads to expensive failures. Plan for 1-year milestones within a 3-year vision.

Your 48-Hour Challenge

Before moving to Module 10, complete at least one of these exercises:

1. **Map your current AI investments:** For each AI tool or initiative, categorize it as Commodity Efficiency, Process Advantage, Data Advantage, or Network Advantage. What does this distribution tell you about your strategy?
2. **Conduct a data asset inventory:** Identify three to five unique data assets your organization possesses that could power AI advantage. For each, assess: how accessible is it, how could it compound value, what would it take to use it?
3. **Draft your 1-year roadmap:** Using the framework in this module, sketch out a four-quarter AI roadmap for your organization. What happens in each quarter? What are the success metrics?
4. **Identify one differentiation opportunity:** Where could AI create a customer experience advantage that your competitors would struggle to match? What would it take to build this?

What's Next

By this point in the book, you might be feeling a bit overwhelmed. That's normal. You've absorbed a lot: frameworks for evaluation, implementation timelines, governance policies, ROI calculations, change management, competitive strategy. It's a lot to hold in your head.

Module 10 is the antidote. It's about building a sustainable system for staying current with AI that takes about 30 minutes a month, not 30 hours. Because the worst thing you can do after finishing this book is trying to follow every AI development and burn out.

THE AI OWNER'S MANUAL

Staying Current Without Losing Your Mind

A sustainable system for staying informed about AI developments without it becoming a second full-time job.

Module Overview

Congratulations, you've made it to the final module of this part of the course. You've built a foundation in AI strategy, learned how to identify opportunities, implemented governance frameworks, calculated ROI, navigated the people side of change, and developed a strategic roadmap for competitive advantage.

But here's the uncomfortable reality that can undermine everything you've learned:

the AI space will look different tomorrow than it does today.

New tools emerge weekly. Capabilities that were impossible become routine. Pricing models shift. Best practices evolve. The "right" answer in January may be outdated by July. For busy executives who already struggle to find time for their core responsibilities, keeping up with AI can feel like an impossible additional burden.

This module addresses that challenge head-on. We won't try to predict the future of AI. Anyone who claims certainty about where things are heading is selling something. Instead, we'll build sustainable systems for staying informed, frameworks for evaluating new developments, and habits that take minutes rather than hours.

The goal isn't to become an AI expert who tracks every announcement. It's to develop the judgment to know what matters for your business, the discipline to ignore what doesn't, and the network to tap when you need deeper expertise.

Learning Objectives

- ✓ Distinguish between AI news that requires attention and noise that can be safely ignored
- ✓ Build a curated information diet from high-quality sources appropriate for executive decision-making
- ✓ Implement a monthly AI review habit that takes 15-30 minutes
- ✓ Apply the When to Adopt Framework for evaluating new AI capabilities
- ✓ Develop an AI advisory network combining internal champions and external expertise
- ✓ Create sustainable systems for organizational AI learning that don't depend on you alone

The Signal vs. Noise Problem

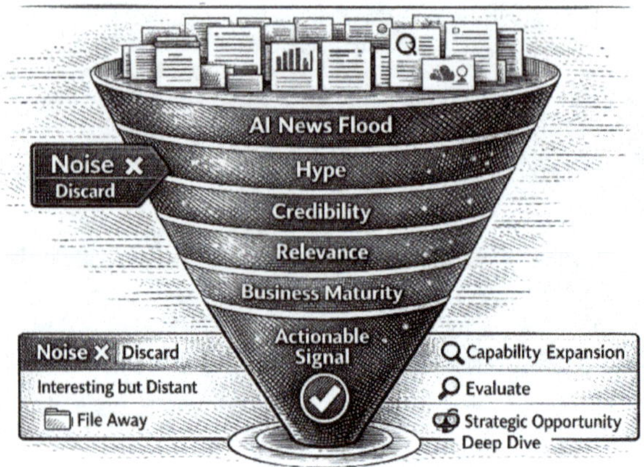

The volume of AI news, announcements, and opinions is overwhelming. Every day brings new product launches, research papers, funding announcements, and hot takes. Social media amplifies everything, making minor updates feel like paradigm shifts and genuine breakthroughs hard to distinguish from marketing hype.

For executives, this creates a real problem. Pay too little attention and you might miss a development that genuinely affects your business or industry. Pay too much attention and you'll drown in information while actual work suffers.

Why AI News Is Particularly Noisy

Several factors make AI uniquely difficult to follow compared to other business technology trends:

Hype incentives. Venture capital, media attention, and stock prices all reward bold claims about AI capabilities. Companies have strong incentives to overstate what their technology can do. Researchers compete for attention in a crowded field. Journalists know that dramatic AI headlines generate clicks.

Capability confusion. Demonstrations of what AI *can* do in controlled settings often differ significantly from what it *reliably* does in production. A model that writes impressive code in a demo may produce errors in real-world use cases. The gap between "possible" and "practical" is often vast.

Rapid iteration. Major AI providers release updates weekly or even daily. Each update brings new features, changed capabilities, and adjusted pricing. Keeping track of what changed, what it means, and whether it matters is genuinely difficult.

Echo chambers. Social media algorithms serve content that confirms existing beliefs. If you're AI-enthusiastic, you'll see more breathless hype. If you're skeptical, you'll see more criticism. Neither perspective gives you accurate signal about business impact.

Missing context. Most AI news lacks the business context executives need. A technical breakthrough may or may not affect your industry. A new tool may or may not solve problems you actually have. Without that context, it's impossible to evaluate importance.

The AI News Evaluation Framework

Not all AI news deserves equal attention. This framework helps you quickly categorize developments and decide how much mental energy to invest.

Category	Description	Action Required
Strategic Shift	Fundamental changes affecting entire industries or business models	Deep analysis, leadership discussion, possible strategy revision
Capability Expansion	New features or abilities in existing tools that could improve current workflows	Evaluate fit, test if promising, update processes if valuable
Competitive Intelligence	How peers or competitors are using AI	Note for reference, evaluate if relevant to your situation
Interesting But Distant	Impressive developments not yet applicable to your business context	File away, revisit in 6-12 months
Noise	Marketing hype, speculation, drama, or developments irrelevant to business use cases	Ignore completely

The key insight is that most AI news falls into the last two categories: interesting but not immediately relevant, or pure noise. Strategic shifts are rare, perhaps a few per year. Capability expansions worth evaluating might number a few per quarter. Learning to quickly categorize and move on is the most important skill for sustainable AI awareness.

Red Flags: News to Approach with Skepticism

Certain patterns in AI coverage almost always indicate hype over substance. When you see these red flags, your default should be skepticism:

- **"Revolutionary" or "game-changing" language:** Genuine breakthroughs are usually described more carefully by those who understand them
- **Demos without production examples:** If you only see controlled demonstrations and no real-world case studies, the technology isn't proven
- **Claims without limitations:** Every AI technology has significant constraints; if they're not mentioned, you're not getting the full picture
- **Vendor announcements without independent validation:** Companies describing their own products are not objective sources
- **Timeline predictions:** "AI will replace X within Y years" claims have an extremely poor track record
- **Social media virality:** What spreads online often has an inverse relationship to business relevance

Building Your Information Diet

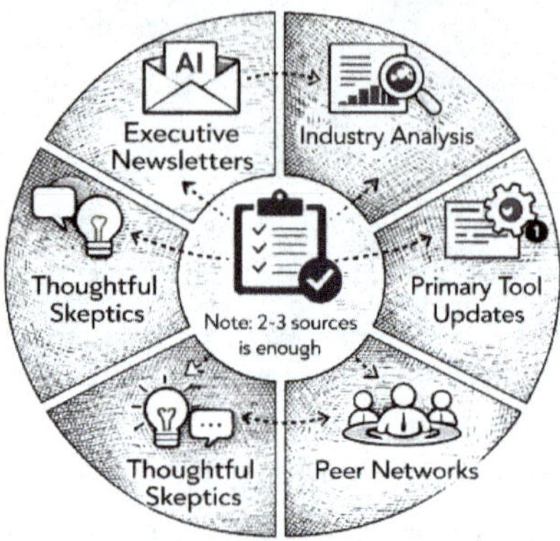

Rather than trying to monitor everything, successful executives curate a small set of high-quality sources that provide signal without overwhelming noise. The goal is a sustainable information diet. Enough information to stay informed, not so much that it consumes excessive time.

Recommended Source Categories

A balanced AI information diet includes several types of sources, each serving a different purpose:

Executive-Focused Newsletters provide curated weekly or monthly summaries written for business leaders. They filter the noise and translate technical developments into business implications. Examples include newsletters from major consulting firms, business publications, and curated AI briefings. One or two well-chosen newsletters can replace hours of browsing.

Industry-Specific Analysis covers how AI is affecting your particular sector. Trade publications, industry associations, and analyst reports often provide more relevant context than general AI news. The question "How is AI changing healthcare (or manufacturing, or financial services)?" is more actionable than "What's new in AI?"

Primary Sources include release notes and blogs from the AI tools you actually use. When OpenAI, Anthropic, Microsoft, or Google announce changes, those directly affect your operations. You don't need to follow every AI company, just the ones whose tools you've deployed.

Peer Networks remain one of the most valuable sources. Vistage groups, industry associations, executive peer groups, and similar forums let you learn from others who face similar challenges. "What's actually working for other companies like mine?" is often the most valuable question.

Thoughtful Skeptics provide balance. Following at least one voice that critically examines AI claims helps counter the enthusiasm bias in most coverage. Good skeptics don't dismiss AI, they demand evidence and point out limitations.

Sources to Avoid or Limit

Equally important is knowing what to skip:

- **Social media AI influencers:** Their incentives (engagement, followers) rarely align with providing accurate business guidance and are usually aligned to affiliate marketing or sponsorships
- **Breaking AI news:** Let stories develop for a few days before paying attention; initial reports are often wrong or missing context
- **Technical research papers:** Unless you have deep technical expertise, wait for expert interpretation rather than reading primary research
- **Vendor content:** Useful for learning about specific products you're evaluating, but not for objective AI education

The Monthly AI Review Habit

Rather than trying to monitor AI developments continuously, most executives benefit from a concentrated monthly review. This approach provides regular updates without daily distraction. The key is making it a scheduled habit rather than something that happens "when there's time."

The 30-Minute Monthly AI Review

Block 30 minutes on your calendar each month. The same day works best (e.g., first Monday of each month). During this focused time, work through the following checklist:

Review your curated sources (10 minutes)

- Scan your executive newsletter(s) from the past month
- Check release notes from AI tools you use
- Note any developments that fall into the "Strategic Shift" or "Capability Expansion" categories

Assess current initiatives (10 minutes)

- Check in on active AI projects: Are they on track? Any blockers?
- Review usage and adoption metrics for deployed tools
- Identify any new opportunities or problems that have emerged

Decide and document (10 minutes)

- Is anything important enough to discuss with your leadership team?
- Are there new capabilities worth testing or evaluating?
- Should any current initiatives be accelerated, adjusted, or paused?
- Document any decisions or action items

Quarterly Deep Dives

In addition to monthly reviews, schedule a quarterly session (60-90 minutes) for deeper reflection:

- Review your AI roadmap progress against plan
- Assess whether your current AI investments are delivering expected value
- Evaluate whether any new capabilities have emerged that warrant strategy adjustments
- Discuss competitive field: How are others in your industry evolving?
- Update your AI roadmap for the next quarter based on learnings

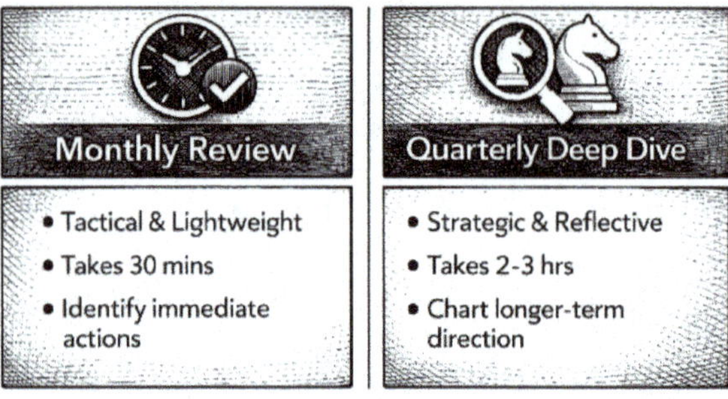

When to Adopt vs. When to Wait

The pressure to adopt new AI capabilities can feel relentless. Every announcement seems to suggest you're falling behind if you don't act immediately. But rushing to adopt immature technologies wastes resources, while waiting too long cedes advantage to competitors. The challenge is knowing when a development is ready for your business and not when it makes headlines.

The Adoption Readiness Assessment

When evaluating whether to adopt a new AI capability, assess these five factors:

Maturity. How long has this capability been available in production? Technologies that have been deployed for 6+ months with documented case studies are significantly less risky than brand-new announcements. Ask: "Who else is using this successfully, and can I talk to them?"

Fit. How well does this solve a problem you actually have? A remarkable technology that addresses issues you don't face is a distraction, not an opportunity. Ask: "If this worked perfectly, how much would it matter to our business?"

Integration. How does this fit with your existing systems and processes? Even excellent tools create friction if they don't integrate with your current workflow. Ask: "What would it take to actually implement this in our environment?"

Resources. Do you have the capacity to implement and support this? Time, money, and attention are finite. Adding a new AI initiative means either dropping something else or stretching already-busy teams. Ask: "What won't we do if we do this?"

Risk. What's the downside if this doesn't work? Some experiments are cheap to run and easy to reverse. Others involve significant investment, data migration, or organizational change. Ask: "How much does it cost to be wrong?"

The Adoption Decision Matrix

Based on your assessment, use this matrix to guide timing decisions:

Scenario	Indicators	Recommended Action
Move Now	High maturity, strong fit, manageable integration, available resources, limited risk	Begin implementation planning immediately
Pilot Test	Moderate maturity, good fit, some integration questions, limited scope possible	Run a bounded pilot to test assumptions before broader rollout
Watch and Wait	Low maturity, unclear fit, complex integration, or constrained resources	Monitor development, revisit in 3-6 months
Skip Entirely	Poor fit, no clear problem it solves, or higher-priority initiatives competing for resources	Remove from consideration, redirect attention

Remember: The courage to skip is underrated.

Most AI capabilities won't be relevant to your specific situation. Saying "this isn't for us" is often the right answer, and it frees resources for initiatives that actually matter.

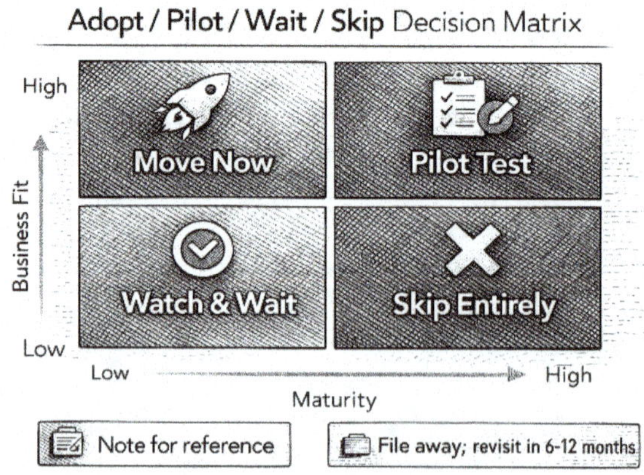

When Early Adoption Makes Sense

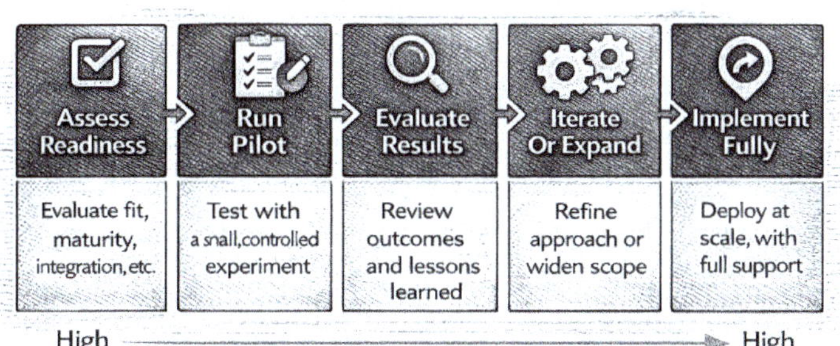

While caution is generally wise, some situations warrant moving faster than competitors:

- **Proprietary data advantage:** If you have unique data that could create compounding value (as discussed in Module 9), earlier adoption lets that flywheel start spinning sooner
- **Customer expectation shifts:** When customers begin expecting AI-powered experiences from your industry, lagging becomes risky
- **Organizational learning:** Sometimes the value isn't the specific technology but what your organization learns from implementing it
- **Competitive pressure:** When competitors are actively deploying similar capabilities, waiting too long cedes advantage

Building Your AI Advisory Network

No individual can maintain expertise across the rapidly evolving AI market. Successful executives build networks that provide access to knowledge when it's needed, without requiring them to personally track every development.

Internal AI Champions

Within your organization, identify and develop people who can serve as AI resources:

The Enthusiast. Every organization has people who naturally gravitate toward new technology. These individuals often experiment with AI tools on their own and can provide ground-level insight into what's working. Give them license to explore and share learnings.

The Translator. Find someone who can bridge technical and business conversations. Someone who understands enough about AI to evaluate claims and enough about your business to assess relevance. This person is invaluable for filtering news and evaluating opportunities.

The Skeptic. A thoughtful skeptic who asks hard questions about AI initiatives isn't an obstacle, they're an asset. Their pushback helps ensure you're pursuing AI opportunities for the right reasons and with realistic expectations.

Functional Champions. As AI use expands, develop go-to people within each major function (finance, operations, sales, etc.) who understand how AI applies to their domain and can spot relevant developments.

External Advisory Relationships

Complement internal resources with external expertise:

Technology Partners. Vendors and consultants who stay current on AI developments can provide relevant updates without you having to monitor everything. Choose partners who proactively share relevant information rather than just responding when asked.

Peer Networks. Executive peer groups like Vistage provide access to leaders facing similar challenges. "What's actually working for companies like ours?" is often more valuable than any expert analysis.

Industry Associations. Many industry groups now provide AI-focused programming and can connect you with peers in your specific sector who are further along in AI adoption.

Advisory Boards. If you have a formal advisory board, consider whether AI expertise should be represented. For critical strategic decisions about AI direction, outside perspective is valuable.

Cultivating Your Network

A network only works if relationships are maintained. Consider these practices:

- Schedule regular (quarterly) conversations with key advisors
- Share what you're learning, networks are reciprocal
- Ask specific questions rather than general "what's new?" updates
- Create internal channels (Slack, Teams) where AI learnings can be shared across your organization

Creating Organizational Learning Systems

Sustainable AI awareness can't depend solely on executive attention. The most resilient approach distributes learning across the organization so that relevant developments surface through normal operations.

Embedding AI Awareness into Existing Processes

Rather than creating new AI-focused meetings and reports, integrate AI awareness into processes you already have:

Strategic planning. Include AI market review in annual and quarterly planning cycles. What developments might affect the assumptions underlying your strategy?

Vendor reviews. When meeting with technology vendors, ask about AI-related roadmap items and how their offerings are evolving. This surfaces relevant developments without additional effort.

Team meetings. Add a standing "AI observations" item to relevant team meetings. What are people seeing in their daily work? What questions are emerging? This crowdsources awareness.

Professional development. Encourage team members to include AI skills in their development plans. As they learn, they bring knowledge back to the organization.

Customer conversations. Train customer-facing teams to listen for AI-related needs, concerns, or expectations. Customers often signal market shifts before they appear in industry analysis.

Building a Learning Culture

The organizations that adapt best to AI changes share common cultural characteristics:

- **Experimentation tolerance.** People feel safe trying new tools and approaches, even if they don't always work
- **Knowledge sharing.** Learnings are shared across teams, not hoarded by individuals or departments
- **Healthy skepticism.** Claims are questioned and evidence is valued, but skepticism doesn't become cynicism
- **Continuous improvement mindset.** The assumption is that how we work today will evolve, and that's okay

Key Takeaways

1. **Most AI news is noise.** Strategic shifts are rare; capability expansions worth evaluating are occasional. Learning to quickly categorize and ignore irrelevant developments is essential for sustainable awareness.
2. **Curate your information sources ruthlessly.** A small set of high-quality, executive-focused sources provides more value than trying to monitor everything. Quality over quantity prevents information overload.
3. **Build habits, not heroics.** Monthly reviews and quarterly deep dives create sustainable awareness. Scheduled time beats sporadic attention, and 30 focused minutes beats hours of unfocused browsing.
4. **Know when to wait.** The pressure to adopt every new capability is artificial. Use the Adoption Readiness Assessment to make deliberate decisions about timing rather than reacting to announcements.
5. **Build your network.** Internal champions, external advisors, and peer connections provide access to expertise when needed without requiring you to personally track every development.
6. **Distribute learning across the organization.** Embed AI awareness into existing processes and cultivate a learning culture so that relevant developments surface naturally rather than depending on executive attention alone.

Your 48-Hour Challenge

Complete at least one of these exercises to put this module into action:

1. **Build your information diet:** Identify two to three executive-focused newsletters or sources you'll follow. Subscribe to them, unsubscribe from noise, and schedule your first monthly review on your calendar.
2. **Identify your internal network:** List three people in your organization who could serve as AI champions or resources. Schedule a conversation with at least one of them about their role in keeping the organization current.
3. **Apply the adoption framework:** Take one AI capability you've been considering and run it through the Adoption Readiness Assessment. Make a deliberate decision: Move Now, Pilot Test, Watch and Wait, or Skip Entirely.
4. **Embed learning into a process:** Choose one existing meeting or process where you could add an AI awareness component. Make the change and communicate it to relevant team members.

And a bonus challenge: Review the 48-Hour Challenges from all of Part One. Which have you completed? Which have you been putting off? Pick one you haven't done and schedule time this week to complete it.

Part One Complete: Your Foundation Is Set

You've completed Part One of "The AI Owner's Manual." Over these ten modules, you've built a solid foundation for leading AI adoption in your organization:

- You understand what AI can and can't do today, cutting through hype to focus on practical applications
- You can identify and prioritize AI opportunities that match your specific business context
- You have a framework for launching AI initiatives with appropriate structure and governance
- You know how each executive function can use AI for their specific responsibilities
- You can communicate effectively with AI tools to get much better results
- You have policies and frameworks to protect your organization while using AI
- You can calculate AI costs and ROI with realistic expectations
- You understand the people side of AI adoption and how to bring your team along
- You can build AI capabilities that create lasting competitive advantage
- You have sustainable systems for staying current without overwhelming yourself

This foundation matters. The frameworks, habits, and judgment you've developed will serve you regardless of how specific technologies evolve. You now know how to *think* about AI strategically and that's a capability that won't become obsolete.

But there's a significant gap between "using AI tools" and "building AI into operations." That's where Part Two takes you.

Part One Deliverables Summary

Throughout this course, you've received tools and templates to support implementation. Here's your complete resource library that can be found at
https://jesscoburn.com/executive-ai/ :

Module	Key Deliverables
Module 1: Reality Check	AI Readiness Self-Assessment, 5 AI Quick Wins Checklist
Module 2: AI Triage	AI Opportunity Prioritization Framework
Module 3: First 90 Days	90-Day Implementation Calendar, Pilot Launch Checklist
Module 4: Leadership Team	Executive AI Playbooks (by role)
Module 5: Communicating	Universal Prompt Engineering Guide, Business Prompt Template Library
Module 6: Governance	AI Acceptable Use Policy Template, Data Classification Guide, Vendor Security Checklist
Module 7: AI Economics	AI ROI Calculator, Budget Planning Template, Business Case Template, TCO Comparison Worksheet
Module 8: People Side	AI Change Management Checklist, Employee Communication Templates, AI Skills Assessment
Module 9: AI Advantage	AI Strategic Planning Framework, Competitive Analysis Template, AI Roadmap Planning Template
Module 10: Staying Current	Curated Resource List, Monthly Review Checklist, AI News Evaluation Framework

THE AI OWNER'S MANUAL

Part Two

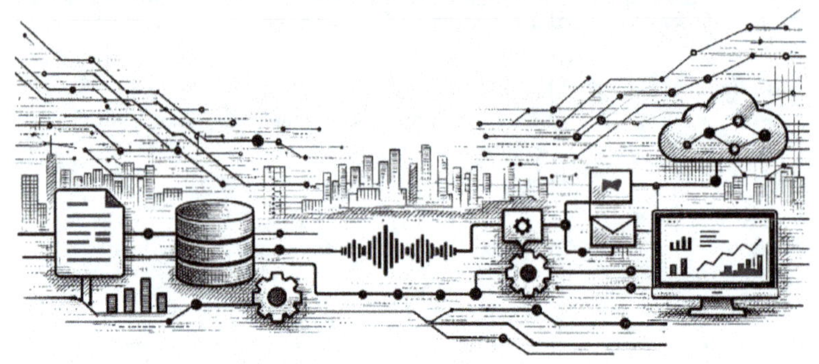

From Foundation to Implementation

Where Competitive Advantage Lives

Part One gave you the foundation: understanding AI capabilities, identifying opportunities, building governance frameworks, managing costs, leading organizational change, and developing strategic advantage. You now think about AI like an informed executive rather than a confused bystander.

Part Two is about building.

Salesforce recently discovered that the gap between "using ChatGPT" and "building AI into operations" is where competitive advantage now lives. 91% of SMBs actively using AI report revenue improvements, yet most remain stuck at experimentation. The organizations pulling ahead aren't just using AI tools, they're weaving AI into their operational fabric.

What Part Two Covers

Five distinct AI implementation patterns have emerged for small and medium businesses. Part Two dedicates a module to each:

Module	Focus	Core Question
11	Advanced AI Implementation	Which implementation patterns are ready for production vs. still overhyped?
12	Automation Platforms	Which platform fits my organization and what will it cost?
13	AI-Enhanced Business Systems	What AI capabilities am I already paying for but not using?
14	AI Voice Assistants	Where does voice AI actually work in business today?
15	RAG and Custom AI	How do I build AI that knows my specific business?

Who Part Two Is For

Part Two assumes you have the foundation from Part One or the equivalent experience. You should be comfortable with:

- The business-problem-first approach to AI evaluation
- Basic prompt engineering principles
- AI governance and data classification concepts
- ROI calculation and realistic cost expectations
- Change management fundamentals

If those concepts feel unfamiliar, the Part One modules will serve you better than jumping straight to implementation.

The Deliverables Ahead

Part Two continues the practical toolkit approach with deliverables including:

- Build vs. Buy Decision Worksheet
- Automation Platform Comparison Worksheet
- Automation Cost Calculator
- AI Business Systems Comparison
- Voice AI Readiness Assessment
- RAG Implementation Planning Guide

Each module maintains the 48-Hour Challenge format with specific actions you can take immediately to apply what you've learned.

Let's build.

THE AI OWNER'S MANUAL

Advanced AI Implementation

Moving Beyond Basic Chat to Operational AI

Module Overview

In the first ten modules, you developed a solid foundation for understanding AI, identifying opportunities, and managing the organizational aspects of AI adoption.

Now it's time to go deeper.

This module marks the beginning of our advanced implementation series. You'll learn about the five distinct AI implementation patterns that have emerged for small and medium businesses: AI agents, enhanced automation, RAG systems, voice AI, and embedded AI in business tools. What matters more: you'll understand which patterns are production-ready versus overhyped, and how to make smart build-versus-buy decisions.

These aren't nice to haves as numbers are compelling: 91% of SMBs actively using AI report revenue improvements, yet most remain stuck at experimentation.

Growing SMBs are 1.6 times more likely than struggling peers to invest in data management before AI tools.

This module will help you join the minority who move from experimentation to industrialized AI and that's where the real financial payoff happens.

Learning Objectives

- ✓ Understand the five AI implementation patterns: agents, automation, RAG, voice, and embedded AI
- ✓ Distinguish between production-ready AI capabilities and overhyped technology
- ✓ Apply the build-versus-buy framework to make smart technology decisions
- ✓ Identify where your organization sits on the AI maturity progression

The Market Has Reached an Inflection Point

The AI space for businesses has matured significantly. What was experimental twelve months ago is now production-ready. What vendors promised "coming soon" is often available today, though frequently not in the form they advertised. Understanding what's real, what's hype, and what's coming next is essential for making smart investments.

Five Emerging AI Implementation Patterns for SMBs

Five distinct AI implementation patterns have emerged, each with different complexity, cost, and readiness levels. Understanding these patterns helps you work through the vendor market and identify the right opportunities for your organization.

AI Agents

AI agents represent systems that plan, use tools, and execute multi-step workflows autonomously unlike chatbots that simply respond to individual queries. These systems can research a topic, draft a report, schedule meetings, and send follow-up communications from a single instruction.

Production-ready options include Microsoft Copilot agents embedded in Office 365, Salesforce Agentforce, and task-specific agents for defined workflows. However, fully autonomous "AutoGPT-style" agents remain experimental with only 11% of organizations actively use agentic AI in production. The gap between demonstration capability and reliable daily operation remains significant.

AI-Enhanced Automation

AI-enhanced automation has evolved beyond simple if/then triggers to intelligent process automation where AI assists decisions within workflows. Traditional automation executed predefined rules; AI-enhanced automation can handle variability, extract meaning from unstructured data, and make judgment calls within defined parameters.

Platforms like Zapier, Make, and n8n now include native AI capabilities for content generation, data extraction, and intelligent routing. This category offers perhaps the best risk-reward ratio for most SMBs. The technology is mature, the costs are reasonable, and the implementation paths are clear.

RAG (Retrieval-Augmented Generation)

RAG grounds LLM responses in your company's proprietary data, reducing hallucinations by 49-67% and making AI relevant to your specific business context. Instead of relying solely on general training data, RAG systems first search your documents, databases, and knowledge bases for relevant information, then use that context to generate grounded responses.

Microsoft Copilot, Salesforce Einstein, and most enterprise AI tools now use RAG architecture. This pattern is particularly powerful for customer service, internal knowledge bases, and any application where AI needs to reference company-specific information accurately.

Voice AI

Voice AI has shifted from frustrating IVR menus to natural language conversations. The technology has matured significantly with 47% of companies now using voice-led technologies, and the market is projected to grow from $5.3 billion to $11.5 billion by 2037.

Production-ready use cases include appointment scheduling (the highest success rate), FAQ handling, order status inquiries, and lead qualification. Complex troubleshooting, emotionally upset customers, and multi-topic conversations remain challenging for voice AI.

Embedded AI in Business Tools

Embedded AI means AI is moving from "add-on feature" to core infrastructure across CRM, accounting, and productivity suites. Google Workspace now includes Gemini in all business plans. QuickBooks and Xero include AI features at no additional cost. HubSpot bundles AI across tiers at reasonable prices.

This pattern requires the least implementation effort. In fact, you may already have AI capabilities you're not using. The first step for many organizations is auditing existing subscriptions for embedded AI features before purchasing new tools.

The Build Versus Buy Framework

The traditional build-versus-buy calculus has flipped for AI. Use AI tools to prototype quickly and understand requirements deeply, then decide whether to buy a solution. This de-risks decisions and often reveals you don't need enterprise solutions.

Approach	When to Choose	SMB Reality
Buy Off-the-Shelf	Commodity use cases, speed critical	70%+ of SMB AI needs
Integrate via APIs	Some customization needed, existing platforms	Best balance for most SMBs
Build Custom	Core competitive advantage, unique data	Rare; requires $100K+ and specialized talent

The key insight: most SMBs should buy off-the-shelf for 70% or more of their AI needs. Custom building should be reserved for capabilities that create genuine competitive advantage, which is rarer than most organizations believe.

Production-Ready Versus Overhyped

Knowing what works reliably today versus what's still maturing prevents wasted investment and failed pilots. Here's an honest assessment based on real-world deployments.

Production-Ready Now

These applications work reliably in production environments today: AI-powered chatbots for Tier 1 support, email and content generation with human review, lead scoring and qualification, document summarization, meeting transcription and action item extraction, basic RAG over company documents, predictive analytics for inventory, and invoice processing.

Still Maturing

These capabilities exist but aren't ready for mission-critical deployment: fully autonomous AI agents that operate without oversight, AI replacing knowledge workers entirely, multi-agent orchestration across complex workflows, AI that "understands" your business without clean data, and one-click enterprise transformation solutions.

The pattern is clear: AI excels at augmenting human work and handling well-defined tasks. It struggles when asked to operate autonomously in ambiguous situations or when the underlying data is messy.

The AI Maturity Progression

Most SMBs are at Stage 1 (experimenting with individual tools) or Stage 2 (building pilots). The financial payoff comes from moving from Stage 2 to Stage 3, where AI is industrialized across multiple processes with consistent governance and measurement.

Stage	Characteristics	Typical Challenge	% of Organizations
Stage 1	Experimenting with individual tools	Scattered adoption, no strategy	~50%
Stage 2	Building pilots with measurement	Scaling successful pilots	~35%
Stage 3	Industrializing across processes	Data infrastructure investment	~8%
Stage 4	AI-native operations	Maintaining competitive edge	~7%

Organizations reaching Stage 3 perform above industry average financially; the 7% at Stage 4 significantly outperform peers. The critical transition from Stage 2 to Stage 3 requires addressing data infrastructure, where most SMBs stall.

Common Mistakes to Avoid

The organizations that fail to capture AI value share common patterns. Avoiding these mistakes really improves your chances of success:

- **Implementing AI without clear strategy:** Technology purchases without business problem alignment waste resources and create shelfware.

- **Ignoring data quality:** 74% of growing SMBs invest in data management versus 47% of declining ones. Data quality is the foundation.

- **Underestimating change management:** Technology implementation is 20% of the work; people adoption is 80%.

- **Starting too big:** Enterprise-wide transformation projects fail. Start with one process, prove value, then expand.

- **Expecting immediate transformation:** AI delivers value through iteration and learning. Plan for 60-90 day breakeven, not instant results.

What Executives Must Decide (Not Delegate)

Some AI decisions require executive judgment and cannot be effectively delegated to technical teams:

1. **AI investment priority:** Where will AI create competitive advantage versus where is it table stakes?
2. **Data strategy:** Who owns data quality? What's the governance model?
3. **Vendor strategy:** Single-ecosystem (Microsoft/Google) versus best-of-breed?
4. **Risk tolerance:** How much AI autonomy before human review is required?
5. **Talent model:** Hire AI specialists versus upskill existing team versus outsource?

What to delegate to technical teams: Specific tool selection within strategic guidelines, technical implementation details, prompt engineering and model tuning, and day-to-day AI operations and monitoring.

Key Takeaways

1. **The gap between experimentation and industrialization is where value lives.** 91% of SMBs actively using AI report revenue improvements, but most remain stuck at Stage 1 or 2. Moving to Stage 3 requires deliberate investment in data infrastructure.

2. **Five patterns, different readiness levels.** AI agents are exciting but still maturing. AI-enhanced automation and embedded AI in business tools offer the best risk-reward for most SMBs today.

3. **Buy before build for most use cases.** 70%+ of SMB AI needs can be met with off-the-shelf solutions. Reserve custom building for genuine competitive differentiators.

4. **Data quality trumps technology selection.** Organizations report 77% rate their data as "average, poor, or very poor" for AI readiness. The best AI tools can't fix bad data.

5. **Some decisions can't be delegated.** Investment priority, data strategy, vendor strategy, risk tolerance, and talent model require executive judgment.

6. **Production-ready is different from demo-ready.** AI-powered chatbots, content generation with review, and document summarization work today. Fully autonomous agents and multi-agent orchestration do not.

Your 48-Hour Challenge

Before moving to Module 12, complete at least one of these exercises:

1. **Audit your existing tools:** Review your current software subscriptions (Microsoft 365, Google Workspace, CRM, accounting) and identify embedded AI features you're not using. Most organizations are leaving value on the table.

2. **Map your AI maturity stage:** Using the four-stage model, honestly assess where your organization sits. What's preventing movement to the next stage?

3. **Identify your #1 data quality issue:** Talk to one department head about data challenges. Is it duplicate records? Inconsistent formatting? Missing fields? Understanding the barrier helps you address it.

What's Next

You now know which implementation patterns are ready for prime time and which ones are still maturing. The natural follow-up question is: OK, but which specific tool do I actually buy?

Module 12 answers that for automation platforms. Fair warning: the pricing on these things is more complicated than the vendors want you to think. I'll show you the hidden costs they don't mention on their pricing pages.

THE AI OWNER'S MANUAL

Automation Platforms Deep Dive

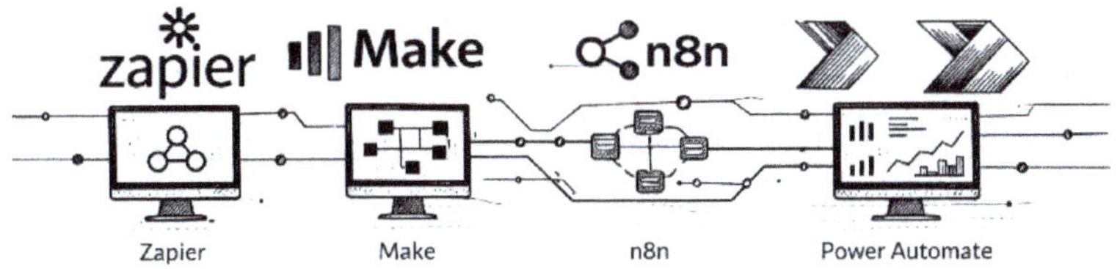

Zapier, Make, n8n, and Power Automate: Choosing the Right Tool for Your Business

Module Overview

In Module 11, you learned about five AI implementation patterns and how to distinguish between production-ready capabilities and overhyped technology. Now it's time to get hands-on with the pattern that offers the best risk-reward ratio for most SMBs: AI-enhanced automation.

Automation platforms have evolved dramatically. What started as simple "if this, then that" triggers has become intelligent process automation where AI assists decisions within workflows. The challenge for executives isn't whether to automate, it's choosing the right platform from an increasingly crowded field and understanding the true costs before commitment.

This module provides the specific, practical guidance you need: real pricing (including hidden costs), honest capability assessments, and a decision framework for selecting the right platform for your organization's profile. By the end, you'll know exactly which platform fits your needs and what to watch out for during implementation.

Learning Objectives

- ✓ Understand the four major automation platforms and their distinct strengths
- ✓ Work Through pricing models and identify hidden costs before they impact your budget
- ✓ Select the right platform based on your organization's technical capacity and use cases
- ✓ Distinguish between automation capabilities that work reliably versus marketing claims

The Automation Platform Overview

Four major platforms dominate the automation space, each with different strengths and ideal use cases. Understanding these differences prevents costly platform switches after you've invested in building workflows.

Zapier: The Easy Entry Point

Zapier remains the easiest entry point with over **8,000+ integrations** and true no-code usability. If you can describe what you want to accomplish, Zapier can probably do it, or at least get you 80% of the way there with minimal technical expertise.

AI Features:

- **Copilot:** Describe automations in plain language and let AI build the initial workflow

- **AI by Zapier:** Text generation and summarization built directly into workflows

- **Chatbots:** AI-powered chatbots for embedding on websites

- **Agents:** Autonomous AI assistants that can browse the web, pull live data, and trigger actions across 6,000+ connected apps

Pricing:

- **Free tier:** 100 tasks/month (good for testing, not production)

- **Professional: $19.99/month** (750 tasks)

- **Team**: $69/month (2000 tasks, shared workspaces, more tasks)

- Chatbots add-on: Free basic tier, Pro at $20/month, Advanced at $100/month

- Agents add-on: Free tier with 400 activities/month, Pro with 1,500 activities/month

Critical cost warning: Costs escalate quickly when stacking AI products. **Budget 40-100% more than the base plan if you plan to use AI features**. A "$20/month" automation solution easily becomes $100+/month once you add Chatbots, Agents and task executions on top of a Professional plan. Zapier also charges per task, meaning each action inside a multi-step workflow counts separately. A single automation that touches four apps burns four tasks every time it runs.

Make: Complex Workflows at Lower Per-Execution Cost

Make (formerly Integromat) offers more sophisticated visual workflows and lower per-execution costs, but comes with a critical hidden cost structure that catches many organizations off guard.

The visual scenario builder is more powerful than Zapier's for complex, branching workflows. If you need conditional logic, data transformation, or workflows that adapt based on content, Make handles these scenarios more elegantly.

Pricing:

- **Free tier**: 1,000 operations/month
- **Core: $9/month** (10,000 operations)
- **Pro:** $16+/month (more operations, advanced features)

The hidden cost trap: Make counts each step plus trigger polls as "operations." A 5-step automation running once equals 5 operations. Here's where it gets expensive: polling checks consume operations *even when no data processes*. A simple 3-step automation polling every 5 minutes uses 8,640 operations monthly just for checking before any actual work happens.

When Make wins: Complex workflows with branching logic, organizations with technical staff who can optimize polling schedules, and scenarios where webhook triggers (which don't poll) are available.

n8n: The Technical Team's Choice

n8n offers the most advanced AI capabilities of any automation platform, with approximately **70 LangChain nodes** for building true LLM applications like chatbots with RAG, web-searching agents, and sophisticated AI workflows.

The self-hosted option is the killer feature for cost-conscious organizations. Once deployed on your infrastructure, you pay zero per-task costs regardless of volume. For high-volume automation, this eliminates the unpredictable cost scaling that makes other platforms expensive.

Pricing:

- **Community (self-hosted): Free** with unlimited executions
- **Starter (cloud):** €24/month (2,500 executions)
- Pro (cloud): €60/month (more executions, priority support)

Critical pricing advantage: n8n hosted environments charge per *workflow execution*, not per step. A 75-step workflow running 5,000 times equals 5,000 executions. On Make, that same workflow would consume 375,000 operations. This makes n8n significantly cheaper for complex, high-volume scenarios.

The tradeoff: n8n requires more technical expertise. Self-hosting means managing your own infrastructure. The interface is powerful but less polished than Zapier. This is not the right choice for non-technical teams.

Microsoft Power Automate: The Microsoft Ecosystem Play

Power Automate wins decisively for organizations already deep in the Microsoft ecosystem. The integration with Microsoft 365, SharePoint, Teams, and Dynamics is unmatched, and many organizations already have access through their existing Microsoft licensing.

Key Features:

- **AI Builder:** Form processing and predictions without code
- **RPA Capabilities:** Desktop automation for legacy applications
- **Deep Microsoft Integration:** Native connections to Outlook, SharePoint, Teams, Excel

Pricing:

- **Premium: $15/user/month** (or often **included with Microsoft 365 business licenses** for basic use cases)
- AI Builder add-ons: $500/unit/month (expensive at scale)

Best for: Organizations using Microsoft 365, SharePoint document management, Teams-centric workflows, or needing desktop RPA for legacy Windows applications.

Platform Selection by Use Case

The right platform depends on your organization's profile, technical capacity, and primary use cases. Use this decision framework to guide your selection:

SMB Profile	Recommended	Budget	Technical
Non-technical marketing team	**Zapier**	$50-150	None
Technical ops team, high volume	**n8n (self-hosted)**	$0 (hosting costs only)	DevOps capability
Sales org with Microsoft stack	**Power Automate**	Often included	Basic IT admin
Data-sensitive industries	**n8n or Power Automate**	Variable	IT admin minimum
Customer service automation	**Zapier Chatbots or n8n**	$70-150	Varies by choice
Complex branching workflows	**Make**	$18-50+	Some technical

Newer Entrants Worth Watching

These newer platforms address specific niches that the major players don't serve as well:

Bardeen ($10-15/month) excels at browser-based automation for sales teams. If your team spends significant time on LinkedIn prospecting and lead enrichment, Bardeen automates these specific workflows more elegantly than general-purpose platforms.

Relevance AI ($19-599+/month) focuses on building specialized AI agent "workforces" without coding. Rather than automating simple workflows, Relevance AI creates AI workers that can handle complex, multi-step processes. It's more expensive but targets a different problem than traditional automation.

Relay.app offers human-in-the-loop AI workflows with exceptional usability. The interface is cleaner than most competitors, and the focus on keeping humans appropriately in the loop makes it well-suited for workflows where AI assists rather than replaces human judgment.

What Actually Works Versus Marketing Claims

Honest assessment matters more than vendor promises. Here's what automation platforms actually deliver reliably versus what remains aspirational:

Works Reliably:

- Data syncing between applications (CRM to email, forms to spreadsheets)
- Email automation and intelligent routing
- Lead enrichment with basic AI analysis
- Scheduled reports and notifications
- Form-to-CRM workflows
- Simple FAQ chatbots with defined response paths

Needs Human Oversight:

- AI-generated content (still requires review before publishing)
- Complex multi-step agents (prone to edge case failures)
- Lead scoring and routing (calibration required)
- Customer support escalation decisions

Oversold (Approach with Skepticism):

- Fully autonomous agents "handling everything"
- Zero-touch complex decision making
- Natural language workflow building for sophisticated scenarios

Implementation Guidance

Starting Your First Automation

The most successful automation implementations follow a predictable pattern. Start with a single, well-defined workflow that solves a real pain point before expanding scope.

Week 1-2: Foundation

- Choose one workflow that runs frequently (daily or more)
- Document the current manual process step by step
- Identify triggers, actions, and expected outcomes
- Build the automation in your chosen platform

Week 3-4: Testing

- Run the automation in parallel with manual process
- Compare outputs and identify discrepancies
- Test edge cases: What happens with unusual inputs?
- Set up error notifications so you know when things fail

Month 2: Full Deployment

- Transition to automated workflow as primary
- Document the automation for team knowledge
- Measure time savings and identify next automation candidate
- Establish review cadence for ongoing optimization

Avoiding Common Pitfalls

- **Starting too complex:** Your first automation should be simple enough to debug easily. A 3-step workflow teaches you the platform; a 15-step workflow teaches you frustration.

- **Ignoring error handling:** Every automation fails eventually. Build in notifications and fallback paths from day one.

- **Not documenting:** The person who builds the automation won't maintain it forever. Document the "why" not just the "what."

- **Automating bad processes:** Automation amplifies existing processes. If the manual process is flawed, fix it first.

Key Takeaways

1. **Platform choice depends on organizational profile.** Zapier for non-technical teams prioritizing ease of use. n8n for technical teams needing power and cost control. Power Automate for Microsoft-centric organizations. Make for complex branching workflows.

2. **Hidden costs are real and significant.** Make's operation counting and Zapier's AI add-on pricing can double or triple nominal costs. Calculate true total cost of ownership before committing.

3. **Self-hosting eliminates per-task costs entirely.** For high-volume scenarios with technical capability, n8n self-hosted provides unlimited executions at zero marginal cost.

4. **Start simple, then expand.** Your first automation should be a single workflow solving a real pain point. Build complexity gradually as you learn the platform.

5. **Fully autonomous agents remain oversold.** Data syncing, email automation, and scheduled reports work reliably. Multi-step autonomous agents still need human oversight.

6. **Check existing subscriptions first.** Power Automate is often included in Microsoft 365. You may already have automation capabilities you're not using.

Your 48-Hour Challenge

Before moving to Module 13, complete at least one of these exercises:

1. **Calculate your true automation cost:** Take one workflow you'd like to automate. Estimate monthly task volume. Calculate costs on Zapier, Make, and n8n using the pricing structures in this module. Include AI add-ons if you'd use them.

2. **Test a free tier:** Sign up for Zapier's free tier (or Make's free tier if you're more technical). Build one simple automation. Maybe a form submission to email notification, or new lead to Slack message. Experience the platform before committing.

3. **Audit your Microsoft licensing:** If you're on Microsoft 365, check whether Power Automate is included in your plan. Log in and explore what's available before purchasing additional automation tools.

What's Next

Before you buy any new AI tools, there's something worth checking first. You might already be paying for AI capabilities you're not using. Google Workspace, Microsoft 365, HubSpot, QuickBooks, Xero; they've all added AI features in the last year, many at no extra cost.

Module 13 walks through what's useful in tools you probably already own. It might save you from spending money you don't need to spend.

THE AI OWNER'S MANUAL

AI-Enhanced Business Systems

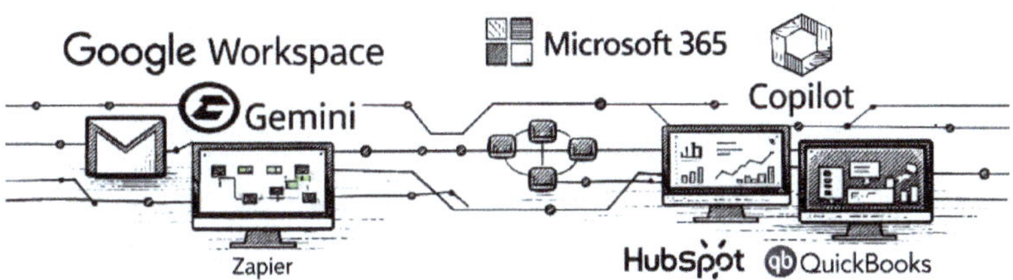

Unlocking the AI Already Embedded in Your Tech Stack

Module Overview

The most valuable AI you can deploy might already be in your software subscriptions. In Module 12, you learned how to evaluate and select automation platforms. Now it's time to examine the AI capabilities embedded in the business tools you're already paying for, and discover which features deliver genuine value versus marketing fluff.

the market has shifted significantly recently. Google Workspace now includes Gemini in all business plans at no additional cost. QuickBooks and Xero have integrated AI features that save SMBs measurable hours each week. Microsoft Copilot has become more accessible to smaller organizations. Meanwhile, some highly marketed AI features remain expensive disappointments.

This module provides the specific, honest guidance you need: real pricing comparisons, capability assessments based on production use, and recommendations for what's actually worth your budget. By the end, you'll know exactly which AI features to activate, which to skip, and how to optimize your tech stack spending.

Learning Objectives

- ✓ Compare the real cost and value of Google Workspace with Gemini versus Microsoft 365 with Copilot
- ✓ Evaluate CRM AI options and understand why HubSpot wins for most SMBs
- ✓ Identify high-value embedded AI features that are included in existing subscriptions
- ✓ Calculate total cost of ownership for AI-enhanced versus traditional tech stacks

Productivity Suite AI: The New Battleground

The productivity suite market has transformed into an AI arms race. Google's decision to include Gemini in all Workspace business plans is a big shift from the previous model of expensive AI add-ons. Microsoft has responded with more accessible Copilot pricing, though the cost structures differ significantly.

Google Workspace with Gemini: Best Value Proposition

Google Workspace now delivers the best value proposition after integrating Gemini into all business plans, with the rollout completing in January 2025. Business Standard at **$14 per user per month** includes full Gemini capabilities in Gmail, Docs, Sheets, Slides, and Meet. With no separate AI add-on required. This is a big shift from the previous $20-30 per month add-on pricing that made AI prohibitively expensive for many SMBs.

What Gemini in Workspace Actually Does:

- **Gmail:** Draft emails from prompts, summarize long threads, suggest replies, improve writing tone

- **Docs:** Generate content, rewrite sections, summarize documents, create outlines

- **Sheets:** Generate formulas from descriptions, create data visualizations, analyze patterns

- **Slides:** Generate presentation drafts, create images, suggest design improvements

- **Meet:** Real-time transcription, meeting summaries with action items, translated captions

Microsoft 365 Copilot: Power at a Premium

Microsoft 365 Copilot remains at **$30 per user per month** as an add-on, requiring a qualifying Microsoft 365 plan (Business Basic at minimum). The previous 300-seat minimum was removed in early 2024, making Copilot accessible to SMBs. A new $21 per user per month tier for organizations under 300 users has been rolling out, though availability varies.

Critical Reality Check on Microsoft Copilot: Adding Copilot to Business Standard ($12.50/user) brings the total per-user cost to $42.50, a **3x increase** over the base subscription. This math matters significantly when budgeting for your organization.

Where Copilot Excels:

- Text summarization across Word, Outlook, and Teams
- Meeting summaries in Teams with action item extraction
- First draft generation in Word and Outlook
- PowerPoint presentation creation from documents or prompts

Where Copilot Struggles:

- **Excel analytics:** Results are non-deterministic and often unreliable for complex data analysis
- **Numerical calculations:** AI struggles with precise math, always verify numbers
- **Brand guidelines:** Generated content rarely matches established brand voice without extensive editing

Head-to-Head Cost Comparison

The cost difference between these platforms is substantial.
For a typical 100-employee organization:

Platform Configuration	Monthly $ (100 Users)
Google Workspace Business Standard (Gemini included)	$1,400
Microsoft 365 Business Standard + Copilot	$4,250
Annual Difference	$34,200 savings with Google

The verdict: For most SMBs, Google Workspace with included Gemini offers the best value. Microsoft 365 with Copilot makes sense only for organizations deeply invested in the Microsoft ecosystem where integration value justifies the premium. Microsoft knows this though and has been adjusting prices to be more competitive.

CRM AI: HubSpot Wins for Most SMBs

The CRM AI market has a clear winner for small and medium businesses, and it's not the enterprise incumbent. HubSpot's AI strategy of bundling capabilities at reasonable prices beats Salesforce Einstein's expensive add-on model for organizations in the 20-500 employee range.

HubSpot AI: Practical Value at Accessible Prices

HubSpot includes its Breeze AI suite: Breeze Copilot, Breeze Agents, and Breeze Intelligence across pricing tiers at reasonable prices. Starter plans at **$20 per seat per month** include meaningful AI capabilities, while Professional at $50 per seat per month unlocks the full suite. The key differentiator: AI features come bundled rather than as expensive add-ons.

What Actually Works in HubSpot AI:

- **Content Assistant:** Genuinely useful for marketing teams drafting emails, blog posts, and social content
- **Email writing:** Generate personalized outreach based on contact data and engagement history
- **Meeting scheduling:** AI-assisted booking that reduces back-and-forth
- **Lead scoring:** Predictive scoring that improves with your data over time

Salesforce Einstein: Generally Not Realistic for SMBs

Honest assessment: Salesforce Einstein is generally **not realistic for most SMBs**. Full AI capabilities run $200-500+ per user per month when including Einstein GPT add-ons.

Implementation costs range $10,000-50,000+. The platform was built on older machine learning technology and requires significant data volume to be effective.

The Salesforce Reality:

- Better suited for organizations with 500+ employees who can justify enterprise pricing

- Requires dedicated Salesforce administrator (often a full-time role)

- AI features require clean, substantial data to deliver value

- Total cost of ownership typically 5-10x HubSpot for comparable functionality

The verdict: Unless you have specific enterprise requirements that only Salesforce can meet, HubSpot's AI capabilities deliver better value for SMBs. The cost difference funds significant other investments.

Accounting AI: Exceptional Value, No Extra Cost

Accounting software represents likely the **best AI value proposition for SMBs** genuinely useful features included in existing subscriptions at no additional cost. Both QuickBooks (Intuit Assist) and Xero (JAX) have integrated AI capabilities that deliver measurable time savings.

What Accounting AI Actually Does

- **Invoice generation from emails/photos:** Snap a receipt or forward an email, and AI extracts the details

- **Automated payment reminders:** AI-timed reminders that adapt to customer payment patterns

- **Receipt processing:** Automatic categorization and data extraction from receipts

- **Transaction matching:** AI suggestions for matching bank transactions to invoices

- **Expense categorization:** Automatic categorization that learns from your corrections

- **Cash flow forecasting:** Predictive insights based on historical patterns

Real Results from Accounting AI

The numbers speak for themselves:

- QuickBooks users get paid **5 days faster** on average with AI invoice reminders

- 45% of users save **12+ hours per month** on bookkeeping tasks

- Receipt processing time drops from minutes per receipt to seconds

The verdict: If you're using QuickBooks or Xero and haven't activated AI features, you're leaving free value on the table. This should be one of your first stops in your AI feature audit.

Collaboration Tools: Meeting AI Is the Clear Winner

Among collaboration AI features, meeting summarization and transcription deliver the most consistent, measurable value. The time savings are immediate and obvious. No longer scrambling to capture notes while participating in discussions.

Zoom AI Companion: Best Free Value

Zoom AI Companion is **included at no additional cost** with paid Zoom accounts. Features include meeting summaries with action items, in-meeting Q&A assistance, and smart chapters for working through recordings. This provides immediate, measurable time savings with zero additional investment.

Slack AI: Valuable for High-Volume Organizations

Slack AI is being bundled into plans, phasing out the previous $10 per month add-on. Pro plans at **$7.25 per user per month** include message summaries and huddle notes. Business+ at $15 per user per month adds enterprise search across connected apps. Valuable for organizations where Slack is the primary communication hub.

Microsoft Teams Copilot: Requires Full Copilot License

Teams AI features require the $30 per user per month Copilot license and they're not available separately. If you're already paying for Copilot, the Teams features (real-time meeting summaries, action item generation, chat thread summaries) are genuinely useful. But the cost only makes sense as part of a full Copilot deployment, not standalone.

What AI Features Are Actually Worth Paying For

After examining the market, here's a practical guide to prioritizing AI feature investments:

High Value: Clear ROI

Feature	Why It's Worth It
Zoom AI Companion	Free with paid Zoom, an immediate time savings
QuickBooks/Xero AI	Included in existing subscription, 12+ hours/month savings
Google Workspace Gemini	Best price-to-value ratio for productivity AI
Canva Magic Studio	Exceptional for marketing teams; design AI that works
HubSpot AI	Good CRM + AI combination at reasonable prices

Conditional Value: Depends on Your Situation

- **Microsoft 365 Copilot:** Only worthwhile for organizations heavily invested in the Microsoft ecosystem
- **Notion AI ($20/user/month):** Good value if Notion is already your knowledge hub
- **Slack AI:** Valuable only for high-volume Slack organizations

Skip for Now

- **Salesforce Einstein:** Too expensive and complex for typical SMBs
- **Multiple overlapping AI subscriptions:** Consolidate before adding more
- **Advanced AI agent features in beta:** Wait until they're production-ready

Total Cost Estimates: Premium vs. Cost-Optimized Stacks

For a typical SMB with 100 employees, the choice of tech stack significantly impacts annual spending. Here's a realistic comparison:

Premium Stack

Component	Monthly Cost
Microsoft 365 Business Standard + Copilot	$4,250
HubSpot Professional	$2,500
Zoom Business	$1,000
QuickBooks (with AI)	$200
Canva Teams	$400
TOTAL (Monthly / Annual)	**~$8,350 / ~$100K**

Cost-Optimized Stack

Component	Monthly Cost
Google Workspace Business Standard (Gemini included)	$1,400
HubSpot Starter	$2,000
Zoom Pro (AI Companion included)	$1,300
QuickBooks (with AI)	$100
Canva Pro (key marketing users only)	$163
TOTAL (Monthly / Annual)	**~$4,963 / ~$60K**

Annual savings with cost-optimized approach: approximately $40,000. This isn't about cutting corners as both stacks deliver thorough AI capabilities. The difference is in ecosystem choice and subscription tier optimization.

Key Takeaways

1. **Audit before you buy.** You likely have AI features in existing subscriptions you're not using. QuickBooks, Xero, Zoom, and Google Workspace all include AI at no additional cost.

2. **Google Workspace with Gemini delivers the best productivity AI value.** At $14 per user per month with AI included, it's 3x cheaper than Microsoft 365 plus Copilot for comparable capabilities.

3. **HubSpot wins for SMB CRM AI.** Reasonable pricing with bundled AI beats Salesforce Einstein's expensive add-on model for organizations in the 20-500 employee range.

4. **Accounting AI is the best free value.** QuickBooks and Xero AI features are included at no additional cost and deliver measurable time savings. 12+ hours per month for many users.

5. **Meeting AI delivers the most consistent collaboration value.** Zoom AI Companion (free with paid accounts) and similar tools provide immediate, obvious time savings.

6. **Stack optimization can save $40K+ annually.** The cost-optimized versus premium tech stack comparison shows that smart choices compound, without sacrificing AI capabilities.

Your 48-Hour Challenge

Before moving to Module 14, complete at least one of these exercises:

4. **Audit your existing AI features:** Log into your QuickBooks/Xero, Google Workspace, and Zoom accounts. Find the AI features you're not currently using. Activate at least one and test it this week.

5. **Calculate your current stack cost:** Add up what you're paying monthly for productivity, CRM, accounting, and collaboration tools. Compare it to the cost-optimized stack in this module. What's the gap?

6. **Test one AI feature for one week:** Pick the AI feature most relevant to your daily work: meeting summaries, email drafting, or invoice processing. Use it consistently for a week and track time saved.

What's Next

Voice AI is the one area where I changed my mind while writing this book. A year ago I would have said it wasn't ready for most small businesses. That's no longer true for certain use cases, though it's still very much not ready for others.

Module 14 sorts out which is which, and the pricing section alone might prevent a costly mistake.

THE AI OWNER'S MANUAL

AI Voice Assistants for Customer Service

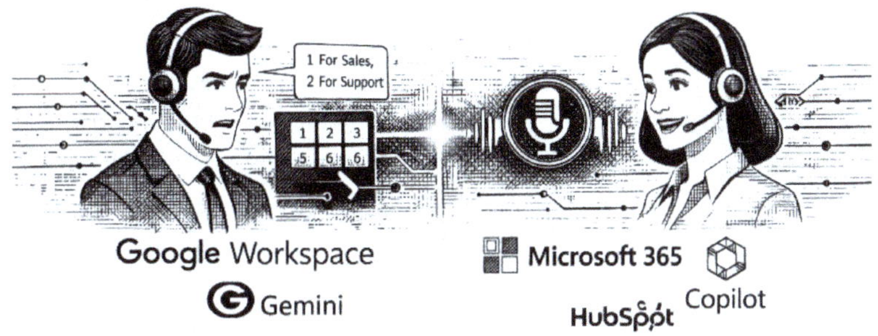

Beyond Frustrating IVR Menus to Natural Language Conversations

Module Overview

Voice AI has reached an inflection point. The technology that once meant frustrating "Press 1 for Sales, Press 2 for Support" menus has evolved into natural language conversations that can genuinely help customers. But as with any maturing technology, there's significant daylight between what vendors promise and what actually works in production.

This module provides the realistic assessment you need to make smart decisions about voice AI. You'll learn which platforms are genuinely production-ready for SMBs, the real costs behind advertised rates, and how to identify use cases where voice AI delivers genuine value versus where it still struggles.

The market opportunity is substantial: 47% of companies now use voice-led technologies, and the market is projected to grow from $5.3 billion to $11.5 billion by 2037. But realizing that opportunity requires understanding both the capabilities and limitations of current technology.

Learning Objectives

- ✓ Understand which voice AI use cases are production-ready versus still maturing
- ✓ Calculate true voice AI costs beyond advertised per-minute rates
- ✓ Select the right platform based on your organization's technical capacity and requirements
- ✓ Work Through regulatory requirements including FCC TCPA compliance

The Technology Has Matured for Specific Use Cases

AI voice assistants have reached production readiness for **structured, predictable interactions**. Common tasks include appointment scheduling, FAQ handling, order status inquiries, and lead qualification. Full customer service automation with complex problem-solving remains challenging.

The key insight: voice AI excels when conversations follow predictable patterns with clear decision trees. It struggles when emotional intelligence, complex troubleshooting, or creative problem-solving is required.

The Pricing Reality: Beyond Advertised Rates

Here's what most vendors don't tell you upfront: platforms advertise $0.05-0.09 per minute rates, but true all-in costs typically range from **$0.13-0.35 per minute** when you include all necessary components.

True Cost Components

Component	Typical Cost
Speech-to-Text (STT)	$0.01-0.02/minute
LLM Processing	$0.03-0.10/minute
Text-to-Speech (TTS)	$0.02-0.05/minute
Telephony/Connectivity	$0.02-0.04/minute
Total Per-Minute Cost	$0.13-0.35/minute

For a typical 4-minute customer call, this means actual costs of $0.52-1.32 versus $3.00-6.00 for human agents. The economics work but only when you understand the true cost structure.

Latency: The Make-or-Break Factor

Latency, the delay between when a customer finishes speaking and when the AI responds, is the single most important factor in voice AI success. Target **sub-800ms latency** for natural conversation flow.

The research is clear: delays greater than one second cause 40% higher hangup rates. Even delays over 500ms cause users to perceive the conversation as unnatural and frustrating.

Platform	Typical Latency
Retell AI	~600ms (best in class)
Twilio ConversationRelay	~491ms (p50)
Bland AI	~800ms
Generic Solutions	950-1500ms (problematic)

Platform Recommendations for SMBs

Not all voice AI platforms are created equal, and most enterprise solutions aren't realistic for SMBs. Here's an honest assessment of what works for organizations with 20-500 employees.

Synthflow: Most SMB-Friendly Option

Synthflow stands out as the most accessible platform for SMBs. It offers a true no-code interface, all-inclusive pricing, and transparent billing. All rare qualities in this space.

- **Starter: $29/month (50 minutes)**: ideal for testing and low-volume use cases
- **Pro: $375/month (2,000 minutes)**: production-ready for most SMB needs
- Pre-built templates for common use cases (appointment scheduling, lead qualification)
- Native integrations with HubSpot, Salesforce, and Zapier

Retell AI: Best Performance, More Technical

Retell AI offers the best latency in the market and strong compliance credentials (HIPAA/SOC 2), but requires more technical expertise than Synthflow.

- Drag-and-drop builder, but more complex than Synthflow
- Best-in-class latency (~600ms)
- 30+ language support
- Typical all-in cost: $0.13-0.31/minute
- Best for: Organizations with some technical capacity who need performance or compliance

What to Skip for Now

Several platforms are either too technical or too expensive for typical SMBs:

- **Vapi and Bland AI:** Require developer resources; too technical for most SMBs
- **Air AI:** Prohibitive enterprise pricing ($25,000-100,000 upfront license fees)
- **Contact Center Platforms (Five9, Genesys, Amazon Connect, Google CCAI):** Generally require 50+ seats or enterprise-level investment
- **Exception: Twilio ConversationRelay ($0.07/minute plus telephony)** works well if you already use Twilio infrastructure

Use Cases: What Works vs. What Struggles

Understanding where voice AI excels versus struggles is critical for setting appropriate expectations and selecting the right initial use cases.

Works Reliably in Production

Use Case	Why It Works
Appointment Scheduling	Highest success rate; structured conversation with clear outcomes
Appointment Reminders	Simple confirmation flow; minimal complexity
FAQ / Information Lookup	Known questions with known answers; RAG enhances accuracy
Order Status	Data lookup with straightforward response
Lead Qualification	Structured qualifying questions; clear escalation path
After-Hours Answering	Better than voicemail; can capture info and schedule callbacks
Payment Reminders	Scripted message with simple response handling
Survey Collection	Structured questions; no complex problem-solving

Struggles Significantly

Use Case	Why It Struggles
Complex Troubleshooting	Requires dynamic problem-solving; too many variables
Emotionally Upset Customers	Requires emotional intelligence AI doesn't have
Negotiation / Exceptions	Needs authority to make judgment calls
Multi-Topic Conversations	Context switching causes confusion and errors
Heavy Accents / Poor Audio	Speech recognition accuracy drops significantly
Medical / Legal Advice	Liability risk; requires human expertise

Regulatory Requirements: What You Need to Know

The February 2024 FCC ruling classified AI-generated voices as "artificial or prerecorded voice" under TCPA. This has significant compliance implications for any organization deploying voice AI.

Core Requirements

1. **Prior Express Consent:** You must obtain consent before making AI voice calls
2. **Disclosure:** Identify the calling entity at the start of each call
3. **Opt-Out Mechanisms:** Provide clear ways for recipients to opt out
4. **AI Disclosure:** Disclosure of AI use is proposed but not yet finalized

State-Level Considerations

Eleven states maintain separate Do Not Call lists with their own requirements. Two-party consent states (including California and Florida) require explicit permission to record calls. If you operate across state lines, you'll need to comply with the strictest applicable standards.

Practical guidance: Consult with legal counsel before deploying voice AI for outbound calling. The regulatory landscape is evolving, and penalties for non-compliance can be significant.

Implementation Framework for SMBs

A phased approach reduces risk and builds organizational confidence. Don't try to deploy thorough voice AI on day one.

Phase 1: Foundation (Weeks 1-4)

- Choose single use case (appointment scheduling recommended as it has the highest success rate)
- Select SMB-friendly platform (Synthflow for most organizations)
- Test with internal calls first
- Target: 100-200 test interactions

Phase 2: Controlled Deployment (Weeks 5-8)

- Deploy for after-hours only initially
- Monitor transcripts and identify failure points
- Tune prompts and conversation flows based on real data
- Target: 500 live calls

Phase 3: Expansion (Months 2-4)

- Add business hours coverage
- Introduce additional use cases
- Integrate with CRM and calendar systems
- Train team on handoff procedures

The ROI Case: When Voice AI Makes Financial Sense

ROI timeline: expect 60-90 day breakeven when replacing routine human calls. Here's the math for a typical scenario:

> **Sample ROI Calculation: 1,000 Calls/Month at 4-Minute Average**
>
> **Voice AI Cost:** $375-520/month (using Synthflow Pro or similar)
>
> **Human Agent Cost:** $4,000-15,000/month (depending on location and benefits)
>
> **Potential Savings:** $3,500-14,500/month

Important caveat: These savings assume the calls are suitable for voice AI (structured, predictable). Attempting to automate complex troubleshooting or emotionally charged interactions often increases costs due to escalations and customer dissatisfaction.

Key Takeaways

1. **Voice AI has matured for structured interactions.** Appointment scheduling, FAQ handling, and lead qualification work reliably. Complex troubleshooting and emotional conversations don't.

2. **True costs are 2-3x advertised rates.** Plan for $0.13-0.35/minute all-in, not the $0.05-0.09 vendors advertise. Budget accordingly.

3. **Latency makes or breaks the experience.** Target sub-800ms response times. Test actual latency, not vendor claims.

4. **Start with Synthflow for most SMBs.** It offers the best balance of usability, features, and transparent pricing. Consider Retell AI if you need better performance or compliance credentials.

5. **Regulatory compliance is non-negotiable.** The FCC's 2024 ruling classifies AI voices under TCPA. Consult legal counsel before deploying outbound voice AI.

6. **Phase your deployment.** Start with after-hours, single use case, internal testing. Build confidence before expanding.

Your 48-Hour Challenge

Before moving to Module 15, complete at least one of these exercises:

1. **Map your call patterns:** Review the last month of customer calls. What percentage are structured/predictable (appointment scheduling, order status) versus complex/emotional? This ratio determines voice AI potential.

2. **Calculate your baseline costs:** What does each customer call currently cost you? Include staff time, training, turnover, and missed calls. Compare to voice AI costs from this module.

3. **Test a platform:** Sign up for Synthflow's free tier or starter plan. Build a simple appointment scheduling flow. Call it yourself. Experience the latency and conversation quality firsthand.

What's Next

Everything we've covered so far, the automation, the business tools, the voice systems, they all use AI in a general way. The AI knows about the world, but it doesn't know about *your* business.

Module 15 changes that. RAG is how you connect AI to your own data so it can actually answer questions about your customers, your products, your internal processes. For a lot of companies, this is where AI goes from "nice productivity boost" to "I can't imagine running the business without it."

THE AI OWNER'S MANUAL

RAG Implementation Making AI Actually Know Your Business

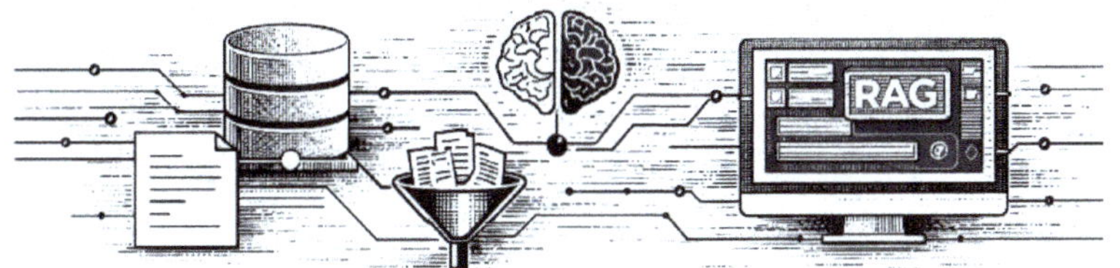

Grounding AI in Your Company's Data for Accurate, Relevant, and Trustworthy Responses

Module Overview

One of the most frustrating limitations of AI assistants is that they don't know anything about your specific business. You can ask ChatGPT brilliant questions about general topics, but ask it about your company's product specifications, your internal policies, or your customer history, and you'll get either generic responses or confident-sounding nonsense.

RAG, Retrieval-Augmented Generation, solves this problem. It's the technology that connects AI models to your company's specific documents, databases, and knowledge bases. Instead of relying solely on general training data, RAG systems first search your data for relevant information, then use that context to generate grounded responses. The result: AI that actually knows your business.

This module cuts through the technical complexity to give you practical guidance on when RAG makes sense, what it actually costs, and how to implement it at different scales. You'll learn to distinguish between the five-figure enterprise platforms and the free-tier solutions that might serve your needs just as well, and understand the trade-offs involved in each approach.

Learning Objectives

- ✓ Understand when RAG makes sense versus using long context windows
- ✓ work through the vector database options and RAG-as-a-service options
- ✓ Estimate realistic costs and timelines for RAG implementation
- ✓ Avoid common pitfalls that derail RAG projects

Understanding When RAG Makes Sense

Not every AI application needs RAG. Modern language models can now handle very long documents directly in their context windows. Claude can process 200,000 tokens (roughly 500 pages) in a single conversation. So when do you need the additional complexity of RAG?

The Decision Framework

The decision comes down to four factors: document volume, user type, update frequency, and audit requirements.

Factor	Long Context Windows	RAG Implementation
Document Volume	Less than 500 pages	More than 500 pages, or growing
User Type	Internal users only	External customers or partners
Update Frequency	Occasional manual uploads	Real-time or frequent updates
Audit Requirements	Informal or none	Citation tracking required
Cost Sensitivity	Low query volume acceptable	High volume, cost-efficient

Key finding from Databricks research: Long context windows cost \$3+ per query at scale, and accuracy degrades after 32K-64K tokens. RAG maintains high accuracy across large document sets with lower per-query costs. For most business applications involving significant document volumes, RAG becomes cost-effective quickly.

A Practical Starting Point

If you're uncertain, start simple. Upload your documents directly to ChatGPT Teams or Claude for Teams. These platforms now support file uploads with enterprise security. Test whether the AI can answer your questions accurately. If it can, and the query volume stays manageable, you may not need RAG at all. If you hit limitations (too many documents, accuracy issues, cost concerns), then evaluate RAG solutions.

Vector Database Options

RAG systems work by converting your documents into "embeddings". Numerical representations that capture meaning. These embeddings live in a vector database, which enables fast similarity searches. When a user asks a question, the system finds the most relevant document chunks and passes them to the AI for generating a response.

The vector database market has matured significantly, with options ranging from free open-source tools to enterprise platforms. Here's what matters for SMB decision-makers.

Managed Solutions (Recommended for Most SMBs)

Pinecone

The most SMB-friendly managed option. Pinecone handles all infrastructure, scales automatically, and includes SOC2/HIPAA certification for compliance-sensitive industries. The free starter tier is generous enough for testing, with Standard plans starting at $50/month. You don't need a database administrator or DevOps expertise. It just works.

Weaviate

Serverless deployment starting at $25/month with a compelling differentiator: built-in hybrid search that combines vector similarity with keyword matching. This often improves accuracy for business documents where exact terminology matters. Open-source with a managed option, giving you flexibility if needs change.

Qdrant

Best performance benchmarks in the category, with a free 1GB cloud cluster and paid tiers from $27-102/month. The trade-off: steeper learning curve and less hand-holding documentation. Best suited for organizations with technical staff who can handle some configuration.

Self-Hosted Options

Chroma

Open source and free when self-hosted. Excellent for prototyping and small-scale applications. You can have a working RAG system in an afternoon. However, it's not production-ready at scale. Use it to test whether RAG solves your problem before committing to paid infrastructure.

pgvector

A free PostgreSQL extension. If you're already running Postgres and have less than 100K vectors, start here. Zero additional infrastructure, familiar tooling, and your existing database team can manage it. The limitation is performance at scale, but most SMB applications never hit those limits.

Quick Comparison

Solution	Starting Price	Best For	Technical Skill	Compliance
Pinecone	Free/$50/mo	Most SMBs	Low	SOC2, HIPAA
Weaviate	$25/mo	Hybrid search	Medium	SOC2
Qdrant	Free/$27/mo	Performance	Higher	SOC2
Chroma	Free (self-host)	Prototyping	Higher	Self-managed
pgvector	Free extension	Postgres shops	Medium	Inherits DB

RAG-as-a-Service Options for SMBs

Building RAG from components (embedding model + vector database + LLM) requires technical expertise. For organizations without dedicated AI/ML staff, RAG-as-a-service platforms bundle everything together. Here are the leading options by ecosystem.

Microsoft Ecosystem

Microsoft Copilot Studio offers no-code agent building with 1,500+ connectors for $200/month (25,000 credits). If you're already a Microsoft 365 shop, this is the path of least resistance. Create chatbots that reference SharePoint documents, answer questions from internal wikis, and integrate with Teams. All without writing code. The limitation: you're locked into Microsoft's ecosystem, and costs can escalate with heavy usage.

Google Ecosystem

Google Vertex AI Search uses pay-as-you-go pricing at $1.50 per 1,000 queries, with 10,000 free monthly queries for testing. Google handles ETL, chunking, and embedding automatically, you just point it at your data. Best for organizations already on Google Cloud Platform who want hands-off infrastructure.

Google NotebookLM is a free tool that deserves special attention for SMBs testing whether RAG solves their problem. Upload PDFs, Google Docs, or web pages, and NotebookLM creates an AI assistant that answers questions with citations to your sources. The standout feature: Audio Overview generates podcast-style summaries of your documents which is genuinely useful for busy executives who prefer listening to reading. Zero cost, zero technical skill required, and surprisingly capable for internal knowledge applications. The limitation: no API or embedding options, so it's purely for direct human use rather than customer-facing applications.

AWS Ecosystem

Amazon Kendra works best for AWS shops with content in supported sources. The pricing model is hourly ($0.32-1.40/hour), which can reach $1,000+/month for always-on deployment. This is enterprise-grade technology, powerful but priced accordingly.

Ecosystem-Agnostic

No-code platforms like CustomGPT, Chatbase, and similar tools offer "ChatGPT with your data" for non-technical users. Upload documents, scrape your website, and deploy an embeddable chatbot, often within hours. Pricing typically runs $50-500/month. These platforms trade sophistication for simplicity. Perfect for customer-facing FAQ bots and internal knowledge bases where advanced features aren't critical.

Realistic Costs and Timelines

RAG implementation costs range from nearly free to six figures depending on your approach. Here's what to actually budget.

Approach	Year 1 Cost	Timeline	Requirements
ChatGPT/Claude Teams + manual uploads	$300-3,000	Days	None - just organize your documents
No-code platform (CustomGPT, etc.)	$1,200-6,500	1-2 weeks	None - point-and-click setup
Copilot Studio + M365 Copilot	$5,000-25,000	2-4 weeks	IT admin capability
DIY with Pinecone + OpenAI	$15,000-50,000	3-6 months	Developer or contractor
Full enterprise platform	$50,000-200,000	3-6 months	ML engineer or SI partner

The honest reality: Most SMBs with 20-50 employees should start in the $300-6,500 range. Organizations with 50-200 employees typically land in the $5,000-25,000 range. Going above that requires either significant technical capability or a systems integrator partnership.

What Accuracy Can You Expect?

RAG really improves AI accuracy for business-specific questions, but it's not magic. Setting realistic expectations prevents disappointment and helps you evaluate whether a deployment is actually working.

Realistic Benchmarks

Well-implemented RAG: 70-90% answer accuracy with citations to source documents. Users can verify where information came from and trust that answers are grounded in actual company data.

Poor implementation: 30-50% accuracy, sometimes worse than vanilla LLM due to retrieval failures. The system finds wrong documents, or correct documents but wrong sections, leading to confidently incorrect answers.

The key factor: Data quality matters more than technology choice. Organizations report 77% rate their data as "average, poor, or very poor" for AI readiness. Clean, well-organized, consistently formatted documents produce much better results than messy data fed through expensive platforms.

Common Pitfalls to Avoid

RAG projects fail in predictable ways. Avoiding these common mistakes really improves your success rate.

Dumping Everything In

The instinct is to upload every document you have. Resist it. More documents means more opportunities for the retrieval system to find irrelevant content. Start with your core documents. The 50-100 files that represent 80% of the questions you want answered. Expand thoughtfully as you learn what works.

Ignoring Metadata

Tags, dates, and categories significantly improve retrieval. A document labeled "2024 Employee Handbook" retrieves better than "handbook_final_v3.pdf" with no context. Invest time in organizing and labeling documents before ingestion. This unsexy work has outsized impact on results.

Skipping Evaluation

Build test cases before deploying. Create a list of 20-30 questions with known correct answers. Run these through your system weekly. Without evaluation, you won't know if changes improve or degrade performance, and performance will drift over time as documents change.

Underestimating Maintenance

Documents change. Policies update. Products evolve. Your RAG system needs to reflect current information, which means establishing processes for keeping the knowledge base current. Budget 2-4 hours per month minimum for maintenance, more if your content changes frequently.

Wrong Platform for Wrong Scale

Enterprise tools for SMB needs waste money. Consumer tools for enterprise needs fail at scale. Match your platform choice to your actual requirements. You'll want to document volume, user count, compliance needs, technical capability.

Don't focus on aspirational future state or vendor promises but your real world need and capabilities.

Recommendations by Company Size

Different sized organizations have different needs, budgets, and technical capabilities. Here's where to start based on your company profile.

20-50 Employees

Start with: Google NotebookLM (free) to test whether RAG solves your problem, or ChatGPT Teams/Claude for Teams with document uploads. Graduate to CustomGPT or similar no-code platform when you hit limitations.

Budget: $0-2,000/year

Technical requirements: None. Someone needs to organize documents and test outputs, but no coding or IT expertise required.

50-200 Employees

Start with: Microsoft Copilot Studio (if M365 shop), NotebookLM or Vertex AI Search (if Google Workspace shop), or CustomGPT. Consider Amazon Kendra if you're on AWS already.

Budget: $5,000-25,000/year

Technical requirements: IT admin capability sufficient. Someone who can configure cloud services, manage connectors, and troubleshoot basic integration issues.

200-500 Employees

Start with: Copilot Studio + M365 Copilot or Vectara for larger document sets. May need custom integration work for complex requirements.

Budget: $25,000-100,000/year

Technical requirements: Part-time developer or systems integrator support. Internal IT can manage day-to-day, but initial implementation likely needs outside expertise.

Key Takeaways

1. **RAG solves a real problem, making AI know your business.** The technology is mature enough for production use, with options at every price point from free to six figures.

2. **Not everyone needs RAG.** If you have fewer than 500 pages of documents and internal users only, long context windows (Claude, GPT) may be sufficient. Test before investing.

3. **Data quality trumps technology choice.** 77% of organizations rate their data as average to poor. Clean, organized documents produce better results than messy data fed through expensive platforms.

4. **Start simple, scale intentionally.** Begin with core documents that answer 80% of questions. Expand thoughtfully as you learn what works and what doesn't.

5. **Match platform to reality.** Most SMBs under 50 employees should spend $500-2,000/year. Enterprise platforms at $50,000+ make sense only for larger organizations with complex requirements.

6. **Plan for maintenance.** RAG systems need ongoing care: documents change, accuracy drifts, and evaluation should be continuous. Budget time, not just money.

Your 48-Hour Challenge

Before moving to the next module, complete at least one of these exercises:

1. **Test long context first:** Upload 5-10 of your most important documents to ChatGPT or Claude. Ask specific questions about them. Can the AI answer accurately? Note where it succeeds and fails. This baseline tells you whether RAG is actually necessary.

2. **Inventory your knowledge base:** Make a list of the documents that would need to be in a RAG system. How many documents? How often do they change? What questions would employees or customers ask? This inventory is the foundation for any implementation decision.

3. **Create 20 test questions:** Write down 20 questions that a perfect AI assistant should be able to answer from your company documents. Include the correct answers. This becomes your evaluation benchmark for any RAG implementation.

What's Next

That's the last of the technical modules. If you've made it this far, you now know more about practical AI implementation than most consultants who charge by the hour to explain it to you.

The conclusion pulls it all together into an action plan. Not more theory. Just: here's what to do first, second, and third.

THE AI OWNER'S MANUAL

Your AI Journey Begins Now

Bringing It All Together and Taking the Next Step

If you've made it this far, you now know more about practical AI implementation than most executives, and certainly more than the vast majority of AI vendors want you to know. You can cut through the hype, identify real opportunities, build a business case, handle governance, manage change, and create sustainable competitive advantage.
But knowledge isn't the same as action. And action is where value lives.

This final chapter synthesizes the key themes from everything we've covered, provides a clear path forward, and because we believe you shouldn't have to figure this out alone, extends an invitation to continue the conversation.

The Seven Truths of AI Implementation

Throughout this book, we've explored countless frameworks, tools, and techniques. But underlying all of them are seven fundamental truths that separate successful AI adopters from the rest:

1. Start with the business problem, not the technology. The executives who fail with AI almost always start by asking "How can we use AI?" The ones who succeed ask "What problems are costing us the most?" and then evaluate whether AI is the right solution.

2. Data quality matters more than model sophistication. The most advanced AI in the world can't overcome poor data. Organizations that invest in data quality before AI tools consistently outperform those chasing the latest technology.

3. Quick wins build momentum; moonshots build graveyards. The path to AI transformation runs through small, measurable successes. Each win builds confidence, skills, and organizational appetite for the next initiative.

4. Governance enables speed; it doesn't slow you down.
Clear policies and guidelines free your team to move faster
because they know the boundaries. Without governance,
every decision requires executive deliberation.

5. Change management is the hardest part. Technical
implementation is rarely what kills AI initiatives. It's the
human factors: fear, resistance, lack of training, poor
communication. Invest in people at least as much as you
invest in technology.

6. AI amplifies; it doesn't replace. The most valuable AI
applications make your best people even better. The worst try
to eliminate human judgment from processes that require it.
Know the difference.

**7. Continuous learning is the only sustainable
advantage.** the AI space will continue evolving faster than
any of us can predict. Organizations that build learning
cultures will adapt; those seeking permanent solutions will fall
behind.

Your 30-60-90 Day Action Plan

Theory is valuable. Action is essential. Here's your roadmap
for the next ninety days:

Days 1-30: Foundation

- **Complete the AI Readiness Self-Assessment**
  (available at https://jesscoburn.com/executive-ai/).
  Know your starting point.
- **Audit your existing tools** for AI features you're already
  paying for but not using.
- **Identify your top three pain points** where AI could
  make a measurable difference.

- **Draft your AI Acceptable Use Policy** (use our template) and get leadership alignment.
- **Start using AI personally** if you haven't already. You can't lead what you don't understand.

Days 31-60: First Win

- **Select one pilot project** using the Process Prioritization Matrix.
- **Define success metrics** before you begin. What will "working" look like?
- **Train your pilot team** on both the tool and the governance framework.
- **Launch and iterate** weekly. Expect to adjust based on real-world feedback.
- **Document everything** for scaling later.

Days 61-90: Scale

- **Measure and communicate results** from your pilot. Numbers matter.
- **Identify two to three adjacent opportunities** based on what you've learned.
- **Recruit AI Champions** from your pilot participants.
- **Plan your organization-wide rollout** and training program.
- **Build your 12-month AI roadmap** connecting AI initiatives to strategic priorities.

You Don't Have to Do This Alone

Throughout this book, we've given you frameworks, templates, and guidance to implement AI independently. And many organizations will do exactly that, successfully.
But we'd be doing you a disservice if we didn't acknowledge a simple truth: having an experienced partner can speed up your journey and help you avoid costly mistakes.

What matters more: we share the philosophy you've encountered throughout this book: business problems first, brutal honesty about what's possible, and practical implementation over theoretical perfection.

How We Can Help

AI Strategy Sessions: A focused conversation about your specific situation, challenges, and opportunities. No sales pitch, just practical guidance.

Implementation Support: Hands-on help selecting tools, configuring systems, training teams, and building governance frameworks.

Managed AI Services: Ongoing partnership to keep your AI capabilities current, secure, and aligned with your evolving business needs.

Training Programs: Customized training for executives, teams, and AI champions based on the frameworks in this book.

A Final Thought

The executives who will define the next decade of business aren't the ones with the biggest budgets or the most advanced technology. They're the ones who understand that AI is a tool. Powerful, yes, but ultimately in service of human goals and human judgment.

They're the ones who start with business problems, not technology solutions. Who invest in their people as much as their platforms. Who build cultures of continuous learning because they know the market will keep changing.

You've invested significant time reading this book. You now have frameworks, tools, and knowledge that put you ahead of most of your peers. But none of it matters until you act.

So here's your final 48-hour challenge:

Pick one thing from this book. Just one. And do it in the next 48 hours.

Complete an assessment. Draft a policy. Try a prompt. Schedule a conversation with your team about AI opportunities. The specific action matters less than the act of starting.

Your AI journey begins now.

Make it count.

— Jess Coburn

CTO & Co-Founder, QIT Solutions

THE AI OWNER'S MANUAL

Supplemental Resources

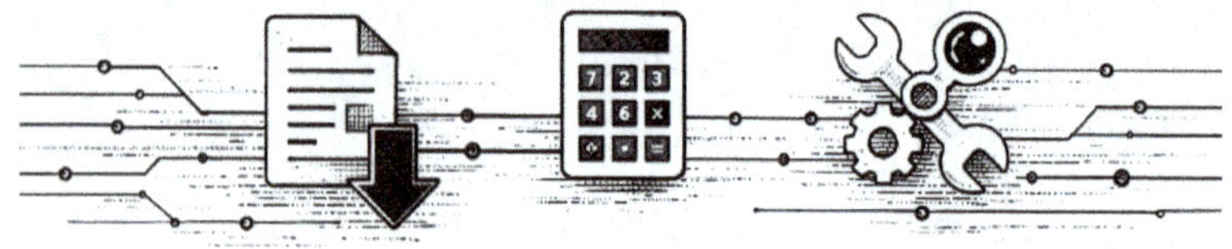

All templates, calculators and tools available at:

www.jesscoburn.com/executive-ai

Module 1: The Executive's AI Reality Check

- **AI Readiness Self-Assessment** - Score your organization's readiness across data maturity, infrastructure, culture, and leadership alignment
- **5 AI Quick Wins Checklist** - Proven, low-risk AI applications you can implement within 48 hours
- **AI Vendor Evaluation Scorecard** - Framework for objectively evaluating vendor claims using the "3 Questions" methodology
- **AI BS Detector Quick Reference** - One-page guide to common AI overselling tactics and red flags

Module 2: AI Triage - Finding Your Starting Point

- **AI Opportunity Assessment Worksheet** - Structured worksheet for identifying and documenting potential AI opportunities
- **Process Prioritization Matrix** - Impact vs. Effort vs. Risk scoring for ranking AI opportunities objectively
- **Industry-Specific Starting Points Guide** - Pre-built opportunity lists by industry to accelerate identification
- **Process Mining Lite Template** - Simple time-tracking template for quantifying current state before AI

Module 3: The First 90 Days

- **90-Day Implementation Calendar** - Week-by-week action plan covering Foundation, First Win, Expansion, and Scale phases
- **Pilot Launch Checklist** - Pre-launch readiness checklist ensuring all prerequisites are met
- **Tool Selection Decision Tree** - Decision framework for selecting AI tools based on use case, budget, and requirements
- **Progress Tracking Dashboard** - Template for tracking implementation progress and adoption metrics
- **First Week Quick Start Guide** - Condensed "just get started" guide for the first 7 days

- **Pilot Retrospective Template** - Structured template for Week 4 retrospective and learnings documentation

Module 4: AI for Your Leadership Team

- **Day in the Life Playbooks** - Role-specific AI integration guides for CEO, CFO, COO, CMO, CHRO, and CTO
- **AI Tools for Business Leaders** - complete guide to AI tools organized by business function and use case

Module 5: Communicating with AI

- **Business Prompt Template Library** - 25+ ready-to-use prompts organized by business function
- **Role-Specific Prompt Libraries** - Customized prompt collections for each executive role
- **Universal AI Prompting Best Practices** - Core techniques that work across all major AI platforms
- **Cross-Platform Prompt Engineering Guide** - Platform-specific optimization for GPT, Claude, and Gemini

Module 6: AI Governance

- **AI Acceptable Use Policy Template** - Customizable policy framework ready for your organization
- **Data Classification Guide for AI** - Framework for determining what data can be used with AI systems
- **Employee AI Training Acknowledgment Form** - Documentation of policy understanding and compliance commitment
- **Shadow AI Discovery Checklist** - Process for identifying unauthorized AI tool usage in your organization
- **AI Incident Response Template** - Step-by-step response procedures for AI-related incidents
- **Vendor Security Assessment Checklist** - Due diligence framework for evaluating AI vendor security practices
- **AI Compliance Matrix by Regulation** - Mapping governance requirements to specific regulatory frameworks
- **AI Governance Board Presentation** - Ready-to-customize presentation for board-level AI governance discussions

Module 7: AI Economics

- **AI ROI Calculator (Excel)** - Multi-worksheet calculator for projecting and tracking AI return on investment
- **AI Budget Planning Template (Excel)** - thorough budget planning with category breakdowns and multi-year projections
- **AI Business Case Template (Excel)** - Framework for building compelling business cases for AI investments
- **AI TCO Comparison Worksheet (Excel)** - Side-by-side total cost of ownership analysis for AI tool comparison
- **AI Hidden Costs Reference Guide** - thorough list of commonly overlooked AI implementation costs
- **AI When NOT to Invest Checklist** - Decision framework for identifying when AI investment isn't warranted

Module 8: The People Side of AI

- **Training Program Outline** - Three-level training structure from AI Literacy to Power User
- **Employee Communication Templates** - Ready-to-customize announcements, updates, and FAQs
- **AI Change Management Checklist** - Phase-by-phase implementation checklist covering all change management activities
- **AI Champion Program Guide** - Complete framework for building an internal AI champion network
- **Manager's AI Conversation Guide** - Scripts and frameworks for discussing AI with team members
- **AI Resistance Quick Reference** - Strategies for different types of AI skeptics
- **AI Skills Assessment for Teams (Excel)** - Evaluate team AI readiness across multiple competency dimensions
- **Role Impact Assessment Worksheet (Excel)** - Task analysis and time reallocation planning for AI-impacted roles
- **AI Adoption Pulse Survey (Excel)** - Track adoption sentiment and progress over time
- **AI Adoption Success Stories** - Curated case studies to help skeptics visualize success

- **Employee AI FAQ** - thorough FAQ addressing common employee concerns

Module 9: Building Your AI Advantage

- **AI Strategic Planning Framework (Excel)** - Develop 1-year and 3-year AI roadmaps with capability assessment
- **AI Competitive Analysis Template (Excel)** - Framework for analyzing competitor AI capabilities
- **AI Roadmap Planning Template (Excel)** - Connect strategic priorities to specific initiatives and timelines
- **AI Data Asset Inventory (Excel)** - Catalog organizational data assets that could enable AI capabilities
- **AI Adoption Readiness Scorecard (Excel)** - Evaluate readiness for new AI capability adoption
- **AI Moat Assessment Framework (Excel)** - Evaluate whether AI creates defensible competitive advantage
- **AI Advisory Network Planner** - Build an external advisory network of AI experts and peers
- **Quarterly AI Deep Dive Template** - Agenda and documentation for quarterly strategic AI reviews
- **Industry-Specific Strategic Playbooks** - Tailored strategic guidance by industry vertical

Module 10: Staying Current Without Losing Your Mind

- **Curated AI Resource List** - Vetted newsletters, podcasts, research sources, and communities
- **Monthly AI Review Checklist** - 15-30 minute structured monthly review process
- **AI News Evaluation Framework** - Quickly categorize AI news as strategic shift, capability expansion, noise, or hype
- **AI Signal vs. Noise Quick Reference** - Pocket reference for daily information filtering
- **Learning Culture Assessment (Excel)** - Evaluate organizational learning culture readiness

Advanced Modules: Implementation Deep Dives

- **AI Implementation Pattern Assessment (Excel)** - Evaluate readiness for agents, automation, RAG, and voice AI

- **AI Maturity Self-Assessment Scorecard (Excel)** - thorough maturity evaluation across all AI dimensions

- **Build vs. Buy Decision Worksheet** - Framework for build, buy, or integrate decisions

- **Data Quality Quick Audit Checklist** - Rapid assessment of data readiness for AI

- **Production-Ready vs. Hype Quick Reference** - Current status guide for AI capabilities

- **Automation Platform Comparison Worksheet (Excel)** - Side-by-side comparison of Zapier, Make, n8n, and Power Automate

- **Automation Cost Calculator (Excel)** - Project true costs including hidden platform expenses

- **First Automation Planning Template** - Step-by-step guide for your first automation project

- **AI Business Systems Comparison (Excel)** - Compare AI features across Google, Microsoft, and standalone tools

- **Embedded AI Features Quick Reference** - Discover AI features in tools you already use

All resources are regularly updated as the AI space evolves.

Check https://jesscoburn.com/executive-ai for the latest versions.

About the Author

Jess Coburn

CTO & Co-Founder, QIT Solutions

Jess Coburn is the Chief Technology Officer and Co-Founder of QIT Solutions, where he drives the company's technology strategy. With nearly three decades of experience in technology leadership, Jess has dedicated his career to translating complex technical capabilities into practical business value for organizations of all sizes.

Throughout his career, Jess has worked with hundreds of business leaders working through technology transformations. From the early days of cloud computing to today's AI revolution. This hands-on experience across diverse industries has given him a unique perspective on what actually works when implementing emerging technologies in real business environments.

Known for his straight-talk approach and ability to cut through vendor hype, Jess believes that technology should solve business problems-not create new ones. His philosophy of "business problem first, technology second" has become a guiding principle at QIT Solutions and runs throughout this book.

www.ingramcontent.com/pod-product-compliance
Lightning Source LLC
Chambersburg PA
CBHW060902140726
47996CB00001B/75